Restoring the Innovative Edge

Innovation and Technology in the World Economy

Editor

Martin Kenney

University of California,

Davis/Berkeley Round Table on the International Economy

Other titles in the series:

Restoring the Innovative Edge

DRIVING THE EVOLUTION OF SCIENCE AND TECHNOLOGY

Jerald Hage

Stanford Business Books
An Imprint of Stanford University Press
Stanford, California

Stanford University Press
Stanford, California

Library of Congress Cataloging-in-Publication Data

Hage, Jerald (Jerald Thore), author.
  Restoring the innovative edge : driving the evolution of science and technology / Jerald Hage.
      pages   cm.—(Innovation and technology in the world economy)
  Includes bibliographical references and index.
  ISBN 978-0-8047-7479-6 (cloth : alk. paper)—
  ISBN 978-0-8047-7480-2 (pbk. : alk. paper)

  1. Technological innovations—United States.   2. Research, Industrial—United States.
I. Title.   II. Series: Innovation and technology in the world economy.
  HC110.T4H34   2011
  658.5′7—dc22   2010043062

Typeset by Westchester Book Group in 10/14 Minion

Dedicated to Madeleine Cottenet Hage, my wife,
for her openness to diverse individuals and countries,
which makes creativity possible, and who has
certainly stimulated mine.

CONTENTS

LIST OF FIGURES AND TABLES

Figures

Tables

ACKNOWLEDGMENTS

IN A BOOK that argues that the atom of innovation is the diverse research team, it is appropriate that I explain how various research teams that I have been involved in helped me to develop the ideas that are expressed in this book. Authors should practice what they preach!

This book had its origin when J. Rogers Hollingsworth (an American historian) called me from Sweden in the middle of the night and asked if I would like to participate in a joint project with him to study the organizational and institutional determinants of scientific breakthroughs in biomedicine. As a consequence of this project, I learned a great deal about the organization of science in American universities from him, and more particularly a National Science Foundation (NSF) grant allowed me to work on the exceptional history of the Pasteur Institute, which is reported in Chapter 4.

One consequence of this collaboration was the organization of a multidisciplinary research team for 1998–99 at the Netherlands Institute for Advanced Studies (NIAS). Because of this collaboration, Marius Meeus (a Dutch sociologist and specialist in studies of innovation) and I organized several international, multidisciplinary conferences with a grant from the Department of Energy (DOE) that produced a book on research agendas in the study of innovation (see Hage and Meeus, 2006). These conferences allowed us to carefully consider definitions for such concepts as knowledge, learning, and innovation as well as to appreciate the complexities of the innovation process. During that year, J. Rogers Hollingsworth and I developed the model of the idea innovation network, which was later published (see Hage and Hollingsworth, 2000). This model led to the evolution-and-failed-evolution thesis reported in Chapter 1.

One of the members of the interdisciplinary team, Frans van Waarden (a Dutch political scientist) organized a four-country European study to test the ideas contained in that model, which is reported in the same chapter. Finally, still another member, Bart Nooteboom (a Dutch economist) exposed us to his ideas about the dilemma between understanding and innovation, which he called novelty, that later became the basis for the solutions to this dilemma expressed in Chapter 3.

In 2001, Gretchen Jordan (an American economist who does evaluation research in science and technology) called me to do some consulting work for her research environment survey, which is reported in Chapter 7. Together we began to develop new evaluation techniques and a framework for conducting multilevel evaluation research (which was published as Hage et al., 2007b, and Jordan et al., 2008). With funding from the Basic Energy Sciences Division of the DOE, the Center for Innovation was able to hire Jonathan Mote (an American sociologist with expertise in network analysis and venture funding of small high-tech companies). With both of these individuals I have conducted a number of studies of national laboratories attached to the DOE and the Center for Satellite Applications and Research (STAR) of the National Oceanographic and Atmospheric Administration (NOAA). This collaboration has produced both reports and published papers that are cited at various points within this book. We continue to work together.

Thus I have learned a great deal from a number of fruitful collaborations, and I wish to thank these individuals for what they taught me. But my biggest debt is to my wife, to whom I have dedicated this book because she has actively participated in bringing it to fruition. Although her own intellectual interests in French literature are far removed from the sociology of innovation, she edited multiple versions of this manuscript as it passed through at least three stages of revision. Her frank and honest criticism, especially regarding repetitious or boring passages, helped greatly to improve the book.

Improvements in this work also resulted from the two anonymous reviewers who critiqued the manuscript for Stanford University Press. For once I found the criticisms thoughtful and penetrating. They gave me a number of good ideas on how to improve the manuscript. Wilbur Hadden, also a member of the Center for Innovation, created the figures for me and provided an enormous amount of assistance with the Notes and the Bibliography. Bart Nooteboom was kind enough to allow me to use his insightful diagram (see Nooteboom, 1999: 14), which has been a considerable spur to my imagination. To each of these individuals, some known and some unknown, I want to express my appreciation.

Finally, I want to express gratitude for my summer home, La Bruyère, perched on a hill in the Cévennes mountains in France where Robert Louis Stevenson took his famous donkey ride. The place is a perfect one in which to think while enjoying the silence only interrupted by the birds. I am very fortunate.

La Bruyère
Arrigas, France

Restoring the Innovative Edge

INTRODUCTION

AMERICAN INDUSTRY faces a severe innovation crisis. Even though the United States had a positive trade balance in eleven high-tech areas, in 2002 the total balance of trade for these sectors went into deficit for the first time in the history of this country (see Figure I.1).[1] In 2008, the high-tech deficit reached $58 billion despite the eroding value of the dollar and the growth in aircraft exports.[2] More recently, however, this strong sector has been threatened by two-year delays in the production of Boeing's new plane, the Dreamliner.[3]

Other indicators point similarly in the direction of a decline in high technology. Several decades ago, the top twenty-five companies ranked according to their RDT (basic research, applied research, and product development) investments were all American companies. Now only nine are, and of these about half were reducing their RDT expenditures between 2003 and 2004—not a positive sign.[4] Not unexpectedly, an annual measure of radical innovations, the top 100 achievements in commercialized products selected by *R&D Magazine*, documents the same kind of decline during the same period. Recognizing that there are limitations in having only 100 awards for achievements in commercialized products and always 100 and that the selection process largely excludes the computer industry and the pharmaceutical industry, the proportion of awards given to the large industrial firms went from about 45 percent in the 1970s to 12 percent in the 1990s, and the downward trend has continued since then with these large firms receiving only six awards in 2006.[5] Still another indicator of technological decline is the proportion of all patents given to the major industrial research firms, especially General Electric (GE), Kodak, AT&T, DuPont, General Motors (GM), Dow Chemical, 3M, United

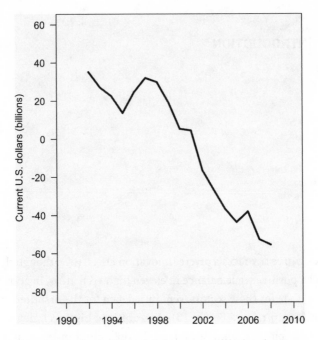

Figure I.1. Decline in U.S. trade balance across all high-tech sectors
Source: Generated from NSF 2010 data.

Technologies, and Ford went from 10 percent to 5 percent between 1972 and 2006 with the biggest drop-off occurring in the 1990s.[6] The concrete manifestations of this decline can be seen in the fading of Bell Labs, Xerox PARC, and other large industrial laboratories that previously produced scientific breakthroughs and technological advances. The major exceptions to these trends remain IBM, Microsoft, Intel, and Sun.

In a parallel manner, the federal government, although it had consistently spent more than 1 percent of gross domestic product (GDP) on RDT during the 1980s, is now spending only .8 percent, just as many foreign countries, especially China and India, are increasing their investments in RDT.[7] Fortunately, President Obama has promised to increase the budget to 3 percent of GDP by 2012. But how much money is allocated to research is only part of the problem; these increases will be achieved only over the years and, given current budget constraints, are anything but certain.

If these negative trade balances are *not* reversed and if American companies and the federal government do *not* invest in the right kinds of research in this country that can restore the innovative edge, then the "rust belts" of the Midwest will be replicated in other parts of the country, particularly along the two coasts (see the pessimistic scenario in the report *Rising Above the Gathering Storm*).[8] The U.S. loss of jobs in the mass production industries of steel, cars, tires, toys, textiles, and others has had a devastating impact on the white and black working-class families resulting in divorce, single parenthood, and one-fifth of American children living in poverty without health insurance. As high-tech jobs are exported overseas and American high-tech industries fail to regain their innovative edge, imagine this same result for white and black middle-class families. Forrester Research estimates that 3.4 million U.S. jobs could be lost to offshoring by 2015, while other economists think that 14 million jobs are at risk.[9]

The United States has already begun to see the signs of this loss with stagnation in the average wages of those in the middle class during the past five years that has accelerated during the current economic crisis as evidenced by the growing deficit in high-tech sectors. The familiar examples of "offshore" middle-class occupations include call center technicians for computer problems, software programmers, and now even researchers. In recent years, IBM has increased its employee base in India from nine thousand to forty-three thousand while laying off thousands of employees in the United States and Europe.[10] So we can only applaud Jeffrey Immelt, the chief executive officer (CEO) of GE, who has said that it is time for American CEOs to rethink outsourcing and start thinking about how to build capabilities in the United States.[11]

The growth of high-tech industries in Asia represents a particular challenge to both the United States and Western Europe. The United States' share of world exports in all high-technology manufacturing declined between 2000 and 2009, specifically in the sectors of communications equipment, office machinery, scientific instruments, pharmaceuticals, and aircraft. Furthermore, it is not just India and China that are players in the global market place. A variety of developing countries are creating dominant niches in particular high-tech markets: Brazil in executive jets, South Korea in dynamic random access memory (DRAM) chips, Finland in cell phones, and Taiwan in boutique chips. Acer, a firm in Taiwan, is poised to overtake Dell as the world's second-largest producer of personal computers, a high-tech area that once was largely owned by American companies.[12] But as a sign that there are also successes in the United States, as this book is going to press, Hewlett-Packard launched a new printer that could receive e-mail messages.[13] Whether this new niche will grow and whether it can be protected by patents remains to be seen, but it does indicate that some American companies remain highly innovative and also how much the success of the United States depends upon being first. Israel's firm Teva now fills more than 600 million prescriptions in the United States, more than Pfizer, Novartis, and Merck combined.[14] Nor should we forget the success of India's pharmaceutical companies in being able to work around the patents held by the American and British pharmaceutical companies. Some of the developed countries are also creating new niches in high-tech areas, such as Denmark in turbines for windmills as an alternative source of energy. France sells metro systems as complete packages, and Germany retains high-quality niches in many products. If the decline in high-tech exports follows the same curve as the steady decrease in the general balance of payments, then we can easily imagine a doubling in the proportion of children living in poverty, in the divorce rate, and in the numbers of homeless in the United States. Therefore, restoring the innovation edge is imperative not only for economic reasons but for social ones as well.

The extent of the innovation crisis can be documented in the increasing number of reports and books that recommended corrective action. First, in December 2004 the Council on Competitiveness' Task Force on the Future of American Innovation recommended that the federal government "increase significantly the research budgets of agencies that support basic research in the physical sciences and engineering, and complete the commitment to double the NSF budget. These increases should strive to ensure that the federal commitment of research to all federal agencies totals one percent of U.S. GDP."[15] Next came the National Academy of Science (Augustine, 2005) assessment, which went further in advocating varied action steps, most of them involving investments in scientific research and education. Concomitantly, various books have appeared with titles like *Innovation Nation* and *Closing the Innovation Gap* with similar appeals for more funding and the training of more scientists and engineers.[16]

While agreeing with these general recommendations, this book is about how to manage the money once it is provided and how the scientists and engineers should be organized. *The key idea is that restoring the innovation edge is not simply one of providing more money or training more scientists and engineers—as important as this is—but understanding what obstacles and blockages are causing the crisis and eliminating them.*

Before discussing these obstacles, perhaps we should recognize one perceptional handicap: Most Americans do not perceive that there is an innovation crisis! In giving talks to various policy groups and in academic settings, I have been amazed at the number of people who do not know the facts presented previously in this Introduction. Many reporters and op-ed writers in our newspapers, magazines, and journals keep saying that the United States is the most innovative country in the world.

Few people perceive that there is a crisis because (1) most people are unaware of the data on the trade balances in high-tech sectors discussed earlier; (2) the recent crises in housing and derivatives as well as the debates about health care reform have pushed the discussion of innovation, which was an important topic from 2004 to 2007, off both the front and op-ed pages as well as from the business section of the major newspapers; (3) paradoxically, a number of local and national success stories in innovation have made it difficult for many to think that we are not innovative enough; and (4) finally, perhaps the most important reason, the concept of productivity and the way in which it is measured have provided us with a false sense of confidence about how well the U.S. economy is performing over the long term as distinct from the current recession. According to this measure, we have the highest per capita GDP, thus leading individuals to believe that we are highly innovative. The problem is that this standard is not a measure of innovation except somewhat indirectly.

Although there are clear signs of a crisis in the data presented at the beginning of this introduction, these data are reported in the National Science Foundation (NSF) *Science and Engineering Indicators*, which is of interest to only a small group of policy makers concerned about science and technology. When I give talks about this problem I have asked the audience how many were aware of the data, and most were not. The major exceptions have been policy makers. The NSF has established a new program of research on the Science of Science and Innovation Policy to develop answers to this problem. In addition, a coordinating committee that unites the evaluation officers of twelve federal agencies has been created to share ideas while the Office of Management and Budget (OMB) has called for research about innovation. It may be a small circle inside the Beltway of the nation's capital, but clearly the experts are worried.

The period 2008–10 with a big housing bubble that burst with drastic consequences for many families, the near financial collapse of the major banks worldwide, and most spectacularly, the rise of unemployment to 10 percent have focused most people's minds on other issues, and yet these issues are clearly related to the innovation crisis. If we are to recapture the lost employment of this past decade, we need to develop innovative products and services that sell on the international markets and provide us with positive trade balances. The important point about the lack of innovation and its impact on unemployment is that this problem has been slowly growing since the 1960s; this recession, with the collapse of the automobile industry, has only made it highly apparent and made it appear to be a temporary situation.

Another reason why people fail to perceive the extent of the crisis is that there are many innovation success stories in the newspapers every month, on both the local level with new start-up companies or in the annual lists of 100 innovative companies found in *Forbes Magazine*, as well as the giants of the computer and Internet industries such as Intel, Apple, Google, Amazon, Facebook, and Microsoft. The highly visible success of first iPhone and now iPad certainly calls attention to Apple's innovativeness. Recently Apple announced the fourth version of its iPhone with a sharper screen and the possibility of video-calling.[17] But as I discuss in Chapter 5, most of the components are made overseas, and thus this innovation does not help American employment. In Chapter 6, I detail a number of areas where the United States invented the technology but then lost control of it because other countries kept improving upon the technology. Later in this Introduction, I analyze the case of robotics. The continued successes in medical research are reported frequently on the various television news programs. Just recently, a new drug for prostate cancer was approved by the Food and Drug Administration (FDA). Yet, we have negative trade balances in both information technology and the health sciences. Thus these reported success stories about innovative products are obscuring the reality of economic failure. And this says nothing about the much larger deficits in the medium- and low-tech sectors where there has been much less innovation, which has been made highly visible with the failure in the automobile industry. As a matter of interest, while the crises of 2008–10 with its impact on the dollar reduced somewhat the extent of the negative trade balances in the low- and medium-tech sectors, it has had no effect on the high-tech sectors.[18] This is further evidence that in these sectors, people do not respond as much to price as to the technological sophistication, quality, and other characteristics. (For a discussion of these issues, see Chapter 1.).

But it is probably the local success stories of start-ups reported in the business sections of our newspapers and made into cases for business students to study that lead the average educated reader to conclude that there is no innovation crisis. Let me provide a few local examples of firms founded in Maryland in the mid-1990s that have become national successes. Honest Tea started when a runner wanted to have good flavored drinks without too much sugar. The company, which was founded in 1996, obtained is first major contract with Whole Foods (Fresh Foods at the time) in 1999. It followed the policy of inventing new tea flavors every year, recognizing that the contemporary consumer has highly varied and customized tastes (see Chapter 1). Because of these successful product launches and Honest Tea's large market niche, Coca-Cola bought a 30 percent interest in the company in 2009.

Another local example in Maryland is Under Armour, which was created by a former University of Maryland football player, at about the same time. Again, it was a similar simple insight, providing athletes with performance-enhancing underwear that did not absorb sweat and kept them cool (i.e., not using cotton). Again, it should be noted that this is a market niche in a highly competitive market dominated by giants such as Nike, Reebok, and Puma. Starting with contracts with major athletic programs, the company has grown rapidly via a policy of product innovation. In 2010 it had almost $1 billion in sales and 10 percent of the high-performance market. These examples of local start-ups can be repeated across many regions of the United States, and their success does make Americans feel that the country is innovative. But again in both cases, it does not mean more manufacturing jobs because the tea and the cloth are imported.

The most important reason, however, why the crisis in innovation is not perceived as such is because of the way in which productivity in the United States is measured. The Department of Labor measures the total number of hours used to produce a unit of goods. By this measure, as the number of work hours declines, the United States becomes more productive. Statisticians at the Department of Labor admit that this standard of measurement is a difficulty. But the problem is that this measure does not include the number of hours of work *outside* the country involved in the production of the same unit of good.[19] Thus if Apple or Under Armour import most of their materials, as they do, they appear to be more productive than they actually are. For example, when earlier tiers in the electronics supply chain, such as semiconductor devices and printed circuit boards, are offshored and these components are imported at lower prices, the remaining downstream domestic industries realize a measured increase in productivity; however, the employment effect is negative.[20] Another problem with this measure of productivity is that it does not include the number of new products or new manufacturing processes that have been developed by American businesses, whether small or large.

But there is another way in which productivity can be misleading if emphasized too much by business elites and policy makers. Productivity measures efficiency or the conservation of resources, whereas measuring innovation is about counting the number of new products and the solving of problems. Thus, a focus on innovation is not only desirable for business growth and creation of jobs, but it is worth emphasizing because it leads to at least partial solutions for the difficult problems that face society. Whether the product is energy-efficient cars, underwear that does not absorb sweat, powerful computers, e-book readers such as the Kindle, or sugarless cookies, or whether the service is a better treatment for breast cancer, more effective screening techniques for terrorists, or a new educational program for the mentally challenged, innovation should be the goal, not efficiency.

At the same time, these two ideas are not always in conflict. One important kind of innovation, new technologies for the production of products or the provision of services, usually reduces the costs involved for any given level of quality. In Chapter 1, I advocate developing a third stage of manufacturing as one way of getting ahead of the innovative curve. In Chapters 5 and 6, I place considerable emphasis on manufacturing because of its implications for employment. In the following section, I discuss the innovation crises from a different perspective.

HOW THE PERSPECTIVE OF THIS BOOK IS DIFFERENT

The major way in which this book is different from current policy discussions is that it focuses on how to manage the innovation process from scientific breakthrough to success in international trade. The focus is always on how to produce radical product and process innovations, both in the private or economic sectors (mainly high-tech industries) and in the noneconomic or public sectors (health, education, homeland security). The public sector is as important as the private sector not only because jobs can be created in these sectors—health has been a major growth industry as its percentage of GDP approaches 16 percent—but because radical innovations help us to extend life, to continue with the health example, improve thinking skills, and reduce risks of terrorist attacks.

Managing innovation involves the following five themes that make this book novel:

1. Coordinating six research arenas in the idea innovation network
2. Recognizing that this coordination has to vary from sector to sector
3. Appreciating that how one coordinates keeps changing because of evolutionary processes
4. Managing innovation at multiple levels, varying from the research teams to science and technology policy
5. Defining management as overcoming obstacles and blockages

When managing the innovation process, what is the objective? The central argument in this book is that restoring the innovative edge requires *radical* product and service program and process innovations. Radical product innovations are products that (1) improve performance significantly (e.g., high-speed trains, hybrid cars, high-definition television [HDTV]) via increased technological sophistication, (2) were previously not available (e.g., e-book readers such as the Kindle, global positioning systems, staining techniques for diagnosing melanoma cancer), (3) represent the subtraction of some undesired quality (e.g., reduction of pollution from manufacturing with scrubbers or from cars with catalytic converters, elimination of sugar and fats in foods), or (4) make the product multifunctional (e.g., optic fibers for TV, Internet, and telephone; tablet computers such as the iPad). In other words, radical product innovations are changes in the dominant design (e.g., shifting from landline telephones to cell phones or from analog to digital representations. Sometimes these are called disruptive technologies because they require learning new skill sets.[21]

Radical process innovations, which is usually what is meant by technological innovation, are defined as manufacturing processes that p (1) significant improvements in the efficiency of the throughput (e.g., a cargo loading and shipping,[22] (2) significant improvement in the qu products produced (reduction of defects and cost of operation), to customize products or services (flexible manufacturing),[23] o reduction in some raw material (such as wood, metal, oil) or energy that is involved in the production process.[24] Whet ucts or radical processes, the basic theme is one of disco previously existing alternatives. This distinction abo product or the technology is an important one bec the emphasis placed on incremental advances in many research teams and firms or the lack of risk taking (see Chapter 2).

These many examples, however, ignore the fact that radical product and process innovations usually require radical changes in their supply and distribution chains. A driver for changes in the supply chain is the development of much more complex products and the use of new kinds of materials to make them. In Chapter 5, I report how important it is to make the supply chain a learning and innovation network as products become more complex, and Chapter 6 reviews ways of revitalizing manufacturing.

But even more important than the supply chain are the idea innovation networks attached to the technologically advanced components in the supply chain. In these networks are six types of research problems that are at least implicit in all radical product and process innovations, even with relatively low-tech products and services. Emphasis on the six kinds of research problems that have to be managed and integrated in what is called the idea innovation network is one distinctive feature of this book (see Table I.1). These arenas reflect the heart of the innovation process at the meso level of economic and noneconomic sectors, the neglected middle between the many books that either emphasizes the societal level of policy or the level of the firm or the research organization. The six arenas are called a network because a good idea may emerge in any one of these arenas and then trigger off research in any one of the others, making the coordination of the six arenas more and more necessary as it becomes more difficult, because the arenas are becoming more differentiated. This evolutionary process is explained at some length in Chapter 1.

The six arenas listed in Table I.1 are familiar ones even though, in the discussion of how to produce a radical product or process innovation, they are not usually mentioned separately, which is an error in our thinking about re-ing the innovative edge. The emphasis in our current thinking has been on T, which more or less corresponds to the first three arenas but ignores er three. These distinctions between arenas of research were not so in the past because all six functional arenas were within the same n such as DuPont, SNCF, GE, or Siemens. But the evolutionary pro-ned in the next chapter are moving more and more of these are-n the largest research organizations whether IBM, AT&T, or of this are provided in Chapters 1 and 5. These processes ex-t toward more "open innovation" upon the part of major

Table I.1. Definitions of the six research arenas in the idea innovation network

Arena	Definition
Basic research / scientific research	Experimental or theoretical work undertaken primarily to acquire new knowledge of the underlying foundations of phenomena and observable facts, without any particular application or use in view.
Applied research / technological research	Original investigation undertaken in order to acquire new knowledge. It is, however, directed primarily toward a specific practical aim or objective.
Product development / product innovation	Systematic work, drawing on existing knowledge gained from research and practical experience, which is directed to producing new materials, products, and devices, including prototypes.
Manufacturing research / process innovation	Research to design new manufacturing products or processes to increase productivity and improve quality.
Quality control research	Research aimed at improving the quality of products as well as research to reduce risks to the user and hidden costs for the environment.
Commercialization research	Research designed to understand needs of customers or to improve distribution channels or find better venues for advertising.

firms with large research and development (R&D) laboratories. However, this evolutionary process varies from one high-tech sector to the next; computers and pharmaceuticals are quite dissimilar.

As the first three arenas—basic research, applied research, and product development (or RDT)—have been discussed many times, they need little amplification. The next three arenas are worth some special attention, especially as they are frequently absent in policy discussions about research and represent some of the problems in our national policies about innovation.[25] Quality research covers a much broader agenda than simply quality control, which is defined as the reduction in defects and operating costs of a particular product. More and more, quality research involves the specification of a number of what might be called "add-ons," that is, the development of a variety of properties that reduce hidden costs for the environment.[26] Examples include reduction in the following negatives: large size (miniaturization), consumption of energy both in the operation of the product and in its manufacturing, and health risks. A particularly striking example of reduction is the strategy of some companies, particularly those with high energy production and consumption, to become "green," such as British Petroleum and GE. In the former case, "green" means developing alternative sources of energy rather than oil. In the latter case, it means not only this but creating products that consume less energy. It also includes the addition of positive qualities such as reductions in noise during operation, ease of disposal, and flexibility in reutilization. From an evolutionary perspective, products are becoming more and complex because of these "add-ons" and reductions in risks to individual users and to the environment.

These objectives pose interesting problems for manufacturing research and for market research. In manufacturing, one has to develop methods for manufacturing various mixes of products defined by their different properties. Some want their food products without sodium, some want it to be cholesterol-free, others search for particular kinds of fats or vitamins, and still others want sugar-free or chemical-free (organic) food. While food products are an extreme example, they reflect the problem of how to manufacture a diverse product mix within the same manufacturing process. In addition to various mixes of products, one needs also to consider the search for more automation and more customization in some of the technological sectors, even automobiles.[27] *One of the themes of this book is that both manufacturing research and quality research have not been emphasized enough, and it is putting us at a disadvantage in world markets.*

Marketing research focuses on the variety of distribution channels that may be necessary for a multifaceted product that comes with many different characteristics. Distribution involves a whole series of questions about supply depots and their location, vital concerns for many Internet companies such as Amazon, while advertising has another set of issues about where, when, and how much to spend to have an impact. The complexity of these issues multiplies as one moves to different countries and considers their cultural context. Entrepreneur programs at business schools emphasize the importance of developing a business plan about how best to distribute a product or service and make people aware of its superior properties. Some of the major success stories in the past decade have been striking innovations in the distribution of services. Examples of this include Starbucks for different kinds of coffee drinks, Netflix for digital video discs (DVDs) and streaming video, and Kindle for downloading e-books.

A key reason for radical changes in the distribution chain is the increasing demand for customization, one of the major evolutionary patterns discussed at some length in the next chapter. Several interesting experiments for providing customization are the Japanese use of salesmen visiting homes to sell automobiles and the Italians using computer measurement to provide customized shoes.

In summary, it is not enough to engage in basic, applied, and product development research to create radical product and process innovations. These have to be integrated with manufacturing research, quality research, and commercialization research before we can expect to restore the innovative edge. Furthermore, problems in either manufacturing or quality research frequently might require solutions in basic or applied research and certainly prototypes typically pose all sorts of problems in manufacturing and quality maintenance. Rather than just thinking in terms of RDT, we need to start emphasizing funding in manufacturing, quality, and commercialization research (basic, applied, product development, manufacturing, quality, and commercialization research—BAPMQC). How to coordinate across these areas is a major focus of Chapters 5 and 6.

Another new feature of this book is the argument that how one manages this network varies from sector to sector. The same prescription does not apply to all kinds of products and services; different coordination mechanisms need to be applied in each sector, which is discussed in Chapters 5 and 6. Beyond this, a substantial amount of economic analysis supports the idea of considerable differences between sectors in one country and even the same sector in different countries.[28]

How different are the high-tech sectors? Figure I.2 only reports the balance of payments in the major ones but they illustrate an important point, namely that the extent of the innovation crisis varies. These sectors differ not only in terms of the amount of RDT invested, to say nothing about manufacturing, quality, and commercialization research, but also the globalization of both their supply chain and their research workers, the number of components in the products, and so on. The nature of competition is distinctive as well. Scientific instruments are usually manufactured in many small high-tech companies, whereas airplanes are concentrated in a few large companies. Much more detail about the differences between the pharmaceutical-biotech sector and the telecommunications sectors is presented in the next chapter. One important conclusion is that scientific policy has to recognize these differences.

The differences between sectors are even greater when the focus shifts to the medium- and low-tech sectors where the greatest job losses have occurred, as explained in the next section. Again, striking contrasts in job losses and the

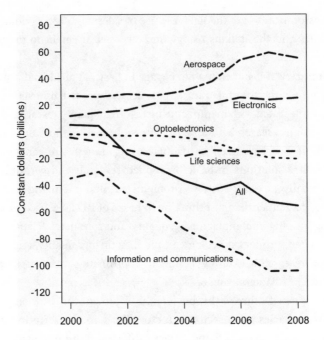

Figure I.2. Selected trade balances for the major high-tech sectors
Source: Generated from NSF 2010 data.

reasons why exist between electric light bulbs (a success story) versus children's toys (a failure), or between clothing (another failure) and chemicals (another success story). *From a managerial perspective, a key difference between sectors is which research arenas are the largest and most important for achieving radical innovations.* One of the arguments made in Chapters 1 and 6 is that more emphasis on manufacturing and quality research can restore our low- and medium-tech sectors.

Still a third distinctive feature of this book is the recognition that these differences hardly remain static but actually grow with time! Except for the theme of globalization and the rising importance of China and India, the general policy and social science literatures, with a few exceptions, have failed to recognize that how one creates radical product and process innovations is continually changing and evolving, *but in predictable directions.* Except for a few sectors, these processes explain the movement toward more "open innovation," again a difference between them. The innovation process is not contained within the single laboratory of a large industrial firm (e.g., aircraft construction) or a national research laboratory (e.g., research on the stockpiling of atomic weapons) but is evolving toward ever more complex arrangements.[29] Indeed, the continued launching of radical products or programs and the technologies to produce them would inevitably have a considerable number of feedbacks that would change the ways in which the next generation of radical innovations would be created.

For managers of research, the major pattern in the evolutionary process is that, more and more, each of the six research arenas becomes differentiated into a separate functional area of research in distinct organizations that must be connected via network coordination before radical product and process innovation can occur. Even the areas of expertise involved in radical innovations change as new research arenas emerge and new scientific and technical training programs are created. As their research arenas become distinct, managers have to be concerned about linking their research team not only with those in other research arenas but those in other sectors of the economy as well. If these evolutionary changes do not occur, the rate of industrial innovation in that sector diminishes. We have one major cause of the innovation crisis in the United States: the failure of managers and organizations to change and evolve in many of these sectors. *Indeed, one of the major reasons that the United States is losing its innovative edge is the failure to adapt to the changes in both the competitive situation and the evolution in the way that new radical innovations are produced.*

Furthermore, the impetus for evolution originates less from the competitive struggle in the market place, however important that is, and more from the growth in scientific and technological expenditures on both national and global levels. Essentially, as more and more money is spent on research, new dominant designs emerge, changing the competitive situation dramatically both nationally and globally.[30] These expenditures also increase the speed with which new products and technologies are developed.

The steady growth in RDT expenditures also means that the development of distinctive organizations specializing in particular kinds of research is occurring now in the low- and medium-tech sectors as they become dependent upon scientific research. Since research is increasingly global in nature, the managers of research organizations and business firms must create an early warning detection system in many geographical places and build bridges to the places where the scientific breakthroughs and technological advances happen. A good example is provided by Procter & Gamble's (P&G's) connect and develop monitoring system that complements their research and development system.[31] This system includes what they call an "Innovationnet," smart learning, and the identification of communities of practice built around their main knowledge communities. Questions are posed about the need for a certain piece of knowledge in the Innovationnet, and sometimes the answers come from surprising places.[32]

Managing the innovation process occurs, at minimum, at four levels, ranging from the research team, where new ideas are processed to the organization in which they are located, to the context of the sector in which competition occurs, to the scientific and technological policies of the nation-state. A comparison of the various books, literatures, and reports indicates that they have honed in on different levels of a complex innovation process.[33] The reality is that all of these levels are part of a complex innovation process that has to be better managed. To restore the innovative edge requires multiple interventions at four distinctive levels: (1) the *research team level* with attention to the amount of diversity and how cross-fertilization is encouraged; (2) the *firm or research organization level* regarding the choice of the kind of public research organization/industrial research organization strategy and transformational team leadership; (3) the *sector level* concerning the choice of how universities, other research organizations, and firms are connected together in teams and networks, and (4) the *societal level* vis-à-vis both industrial and federal governmental support for research and the training of the necessary human capital. The only way in which the four levels can be managed successfully is via cooperation between the public and private sectors, an anathema to many.

The management of innovation, regardless of level and sector, necessitates solving problems or overcoming obstacles. Whenever there is some crisis, it is best to avoid looking for simple solutions or single causes and instead more prudently try and understand what obstacles prevent change. My research and work on innovation leads me to believe that there are at least eight major sets of obstacles, and while I am sure that there are probably more than this, these eight represent a good starting point. My major objective in writing this book has been to help managers and policy makers by suggesting remedies for overcoming these obstacles. The common theme that unites these obstacles is the tendency for individuals, managers of organizations, and policy makers to continue to do things as they have in the past. In the academic literature this is called path dependency. Each obstacle requires the recognition that we cannot continue as we have in the past with static strategies, narrow perspectives in research teams, lack of communication within teams, little coordination between organizations in the same sector, and lack of cooperation between the public and the private sectors, but instead we must change.

What are the sources that have led me to identify these obstacles and their remedies? My first research studies focused on the causes of product and process innovations, and this line of research has continued off and on for more than forty years. In the last two decades, I have concentrated much more on the problems of scientific research. As an example of the latter, some of the insights originated in a decade-old comparative research project on radical innovation in biomedicine.[34] This project focused on the determinants of radical innovation within the research laboratory, the research organization, and the scientific research environment; thus it covered all three levels from the atom of the team, to the molecule of the organization, to the macromolecule of the sector, and to the compound of society, which is composed of many patterns that created constraints.[35] This research led to insights that are reported in Chapter 4. Furthermore, given the increased importance of health research in the United States, this area is another important focal point of this book.

Comparative research provides some of the best opportunities for understanding new ways of restoring the innovative edge. Besides a study of a transformational organization in France, informative examples are taken from Japan, South Korea, Finland, Germany, the Netherlands, Italy, and other nations. But a major comparative learning experience for me was the two international conferences that Marius Meeus and I organized on research problems in innovation, science, and institutional change with a large grant from the Department of Energy (DOE). The many contributors made apparent the need to examine the problem of innovation from multiple levels of analysis and in different sectors.[36]

Other insights about how to manage a complex research process have also benefited from my five years of consulting with a national research laboratory attached to DOE and with the Center for Satellite Applications and Research (STAR) of the National Oceanographic and Atmospheric Agency (NOAA). Working with Gretchen Jordan and Jonathan Mote, we have learned some valuable lessons from case studies and surveys in the large public research laboratories that are reported in this book. In particular, the former has developed a research environment survey that allows for critical feedback on the innovation process (see Chapter 7). The latter has advanced the theory and methodology of studying networks, which have become such a critical coordinating mechanism in the idea innovation network. Together we have developed a complex policy framework for evaluating science and technology so that governments can intervene more intelligently, which informs the identification of obstacles.[37] Therefore, the topics in the book are grounded in a reality, necessary when making recommendations. The issue is then what works and how do we know it works![38] The following section suggests how innovation can help to protect employment.

Given this deep personal concern about the innovation crisis in the United States, I have written this book to try to reach two separate and distinct audiences: academics, my usual audience, and policy makers as well as research managers. For the benefit of the latter, I have attempted to eliminate as much of the academic jargon as possible, relegating much of it to endnotes. Each chapter has a paragraph at the beginning about a basic policy issue that is relevant to the topic being discussed. But for the benefit of the social science academic audience, I conclude each chapter—except this one and the Epilogue—with an explanation for how the discussion of obstacles can contribute to one or more social science literatures. This Introduction argues for a new policy model, one that focuses on both evolution and failed evolution as the basis for policy changes. The Epilogue argues for the need for a new socioeconomic paradigm built upon the concept of innovation. Because this book discusses eight obstacles, it ranges across a number of different specialties and disciplines, and these connections should be drawn as much as possible but in ways that allow those only interested in policy matters to skip over them.

THE ROLE OF INNOVATION IN PROTECTING EMPLOYMENT

Between 2007 and 2009, the U.S. economy lost over 8 million jobs. The assumption may be that many of these jobs will return as we recover from the recession, but the most disturbing aspect is that approximately one-half of these jobs have been lost permanently.[39] One problem with the assumption of a temporary job loss is that many of the 7.5 million jobs created during 2004 and 2007, several years after the high-tech sectors exports became negative, were the result of excessive spending fueled by unsustainable consumer and federal and, in many cases, state government debt rather than the creation of new industries or exports.[40] In other words, the housing bubble had a substantial, negative impact on patterns of consumption. Another problem with the assumption of a temporary job loss is the long-term policies of many, though not all, American manufacturers resulting in the shifting of jobs over-seas, both unskilled and high-tech. A good example of what has happened in employment is the printed circuit boards niche of the semiconductor sector, which had 80,000 workers in 2000.[41] This number dropped almost in half to 41,000 in 2004 and has been steadily declining since then, indicating how quickly skilled employment in high-tech sectors can be lost.

One unfortunate casualty of the current economic crisis has been to dimi-nish the concerns about innovation that were expressed in the period 2002–7, before the current loss of jobs.[42] Even though these alarm bells are relatively recent, the lack of product and especially process innovations in American firms' strategies are deep-seated and demonstrable in a series of choices made in many American industries from the 1960s through to the 1980s, resulting in the acceleration of the total trade deficit during the 1990s and the much larger losses in jobs during the current recession.[43] The CEOs during this time period in general—though there were many exceptions—pursued static strategies.

To understand why, we need to examine the reward system for CEOs that first was based on productivity measures that ignored innovation and then later on stock prices, which encouraged mergers and acquisitions that are even more removed from innovative products and process. This reward system al-lowed for and reinforced the rise in power of, first, accountants, and then, later, finance specialists, who are quite different from managers that come from a research or engineering background.[44] The constant search for reduc-ing costs and increasing return on investment, and the strategic choices that would momentarily inflate stock prices rather than provide a solid foundation for growth have allowed these occupations to shine and receive large bonuses, stock options, and golden parachutes. In other words, the same logic that has led to the excessive greed of Wall Street was also operating prior to this in the executive suites throughout corporate America.

Historians admonish us not to repeat the errors of the past. Therefore, it is worth examining the past four decades for lessons that can inform policy to-day. But we might also argue that we should repeat the successes of the past as well as avoid repeating errors. The economic history of the United States has several positive lessons that have been largely ignored, although some of them are being presently rediscovered. One of the ignored lessons is how much in-novation there has been, not just in the process technologies that provided the productivity and expansion of American manufacturers during the industrial revolution, but more critically in the innovative products and process that drove the American economy during the post–Civil War period until well into the post–Second World War period.[45] But what had been missed in the emphasis on gaining market share via a standardized product with a low price was that frequently the company that become dominant in the United States and also worldwide did so not only with a novel product *and then* process innovations that achieved considerable productivity but pioneered with one or more product innovations: Such companies as Singer Sewing Machine, Diamond Match, Otis Elevators, GE, Westinghouse, DuPont, RCA, AT&T, and GM are only a few examples.[46] Furthermore, many of the more successful companies maintained product innovation over extended time periods, for example, GE, Westinghouse, DuPont, RCA, and even GM, although it is hard to believe it now. In the beginning, these companies followed a policy of product innovation and then emphasized process innovations of various kinds so as to reduce the manufacturing costs. Rather than draw the correct lessons, namely, the need to continuously innovate new products, new processes, and new methods of distribution, the managers of industry in the post–Second World War period since the 1960s focused on the wrong measures of success—bottom lines of corporate profits and stock prices. The correct measures of success for today are growth, the objective of GE, and specializing in radical innovations on the basis of new performance capabilities, as does P&G.[47]

We can gain some insights about the advantages of emphasizing radical product and process innovations for protecting employment by reexamining the reasons for the negative trade balances in our mass production industries such as steel, television, tires, toys, and textiles. By understanding what *could have* at least diminished the number of jobs exported overseas, I can lay a foundation for ideas about how to protect middle-class employment now by in part returning to what made America great prior to the Second World War, that is, the constant emphasis on innovation. This is not to say that the situation today is the one that prevailed in the low- and medium-tech industries from the 1960s through the 1990s. The occupations at risk are different, now mostly middle-class jobs, and the remedy is much more difficult, as is evident in the eight obstacles that have to be overcome to restore the innovative edge. At the same time, the general theme remains largely the same: the critical importance of continuously emphasizing radical innovations in products and processes and the development of new business models for how to compete.

Our review of the past four decades starts with an incorrect assumption about the reasons for the loss of jobs. The general assumption is that the jobs in these industries were located overseas because the workers in other countries could be paid much lower wages. Two alternative strategies would have protected many, if not most, of the low-tech jobs including their wages and benefits. New product innovations would have allowed American manufacturers to move into new markets and slowed down the advantages of low-wage labor elsewhere, while the new process innovations that were available would have considerably increased productivity. New models of how to commercialize products would have made the new products more adapted to the different cultural tastes found in disparate regions and nations.

The best way to understand how the failure to adopt new products resulted in the loss of jobs is to examine a few industrial sectors where this strategy was pursued. A good example is our light bulb industry. One might have assumed that since this product is typical of a mass production product, we would be importing light bulbs from third world countries. However, employment in this sector has been protected by its emphasis on not only radically new products, first halogen bulbs and now light-emitting diodes (LEDs) based on a silicon technology, but also bulbs that consume less energy and last longer, thus reducing operating costs and contributing to the war against global warming. Another positive example was the success of drip-dry shirts during the 1960s. This technology protected part of our textile industry for some time. But unfortunately, it has not been followed up with other radically new products with superior qualities in the textile industry.

Unfortunately, these exceptions are not the general rule. In contrast, when Michelin introduced its technologically sophisticated tires, the American tire manufacturers never thought the American consumer would pay a high price for higher quality and superior performance. As a consequence, most of the American tire industry has disappeared. Another example of a slow response was the failure of Singer Sewing Machine to add electronics to its products in response to Japanese competition. The decline of this company is an exceptionally sad story because it was the first multinational corporation and the first great American success story. Its success was built on a simple innovative product but one that was distributed in a unique way via the Singer Sewing Machine stores where women could be taught how to use the machines and many different kinds of materials for sewing could be purchased.

In contrast, a quick response by Anheuser-Busch to a competitive move of a radically new product, ice beer in California, introduced in the American market by Kiri, a Japanese brewery, protected Anheuser-Busch's market share.

Anheuser-Busch had been carefully monitoring new product development worldwide, was prepared, and launched a competitor within two weeks. The historical lesson is that one has to monitor product development globally and be prepared to respond immediately.

Much of the present argument for the export of jobs relies upon the idea of lower wage costs being available in developing countries such as India and China. But a radical process technology—flexible manufacturing, or what is called post-Fordism—has been available for several decades, and it provides more productivity, higher quality, and most critically, greater flexibility in responding to competitive moves or shifts in tastes.[48] The failure to adopt and successfully implement this new manufacturing technology is perhaps the single biggest cause of the loss of much of American industry in many of the mass production sectors of the American economy.[49] The same failure is typical of many of the Anglo-Saxon economies with their reliance on the business model discussed later.

The principle of flexible manufacturing is that one achieves higher productivity by producing a wide variety of products and even customized ones on the same production line. In 1993, I visited an automobile plant in Japan that was producing trucks, low-priced cars, and medium-priced cars on the same production line. Flexible manufacturing involves being able to reprogram machines so that they can produce a different-sized part for the assembly line.

One of the first American industries that largely disappeared because of the failure to correctly implement flexible manufacturing was footwear. During the 1960s, American shoe machine manufacturers developed automated equipment that would allow for the efficient production of small batches of shoes and quick changes in the styles being produced.[50] However, the engineers controlling the major companies in the footwear industry refused to adopt these machines because their performance measure was productivity rather than innovation, and they felt that a long production run with a single style would produce the highest profits. This is the very same logic that Ford used in its Model T, which is why the first stage in the evolution of manufacturing is called Fordism, or the emphasis on the assembly line with long production runs. Surely, by producing the same product over a number of years, one can reduce the cost per unit a great deal.[51] But, in contrast, the Italian footwear industry, which also has been buffeted by competition from developing countries, has remained largely protected. During the 1980s, the Italian footwear workers were receiving higher wages than their American counterparts, yet their jobs were largely protected on account of their high-quality, well-designed market niches. Admittedly, the success of the Italian footwear industry rested on more than just their adoption of these new flexible machines.[52] Another important factor was the unusual network organization of small firms that appears to facilitate flexibility, the cooperation between workers and managers, the special role of the technical institutes that designed sophisticated machines and trained a highly skilled workforce, and finally, the importance of Italian design.[53] Indeed, this Italian model offers many insights for our community colleges about what they could do for small local businesses by providing the design of new kinds of manufacturing machines, the training of workers to handle these new machines, and of course, the teaching of industrial design, topics that we return to in Chapter 6. The historical lesson from this comparative example is that product and process innovation is not enough to save jobs; it requires other kinds of changes in the environment of the firms as well. Perhaps the best proof that jobs could have been saved in the American footwear industry, for instance, is the experience of the New Balance shoe company, which has maintained about 30 percent of its workforce in the United States with policies of job training.

Nor is the footwear industry the only one in which the Italian model has much to offer, especially in our low-tech agricultural and craft sectors. In the 1940s, American agriculture produced more than 50 percent of the global production of tomatoes. American processors failed to integrate the growing and processing of tomatoes and develop new kinds of tomato sauces including new products (sauces and salsas) as quickly as did Italy and Mexico. Italy has repeated the same successful pattern with new kinds of processing machines in grape growing as well.

To provide an American example of how the adoption of flexible manufacturing can protect employment of working-class jobs, we can turn to the case of washing machines. GE increased the productivity of its manufacturing of dishwashers by 300 percent during the 1980s. Obviously, one can protect a lot of employment and wages with that strong a productivity increase. Interestingly enough, this is the same company that pursued the policy of product innovation that protected the light bulb industry in the United States in the 1990s. The historical lesson that one can draw from this is that low-tech companies can survive when they pursue the correct strategies of adopting and implementing flexible manufacturing techniques coupled with radical product innovation.

Contrary to what one might expect from this example, some of the product innovation pioneers in the United States did not adopt flexible manufacturing. In other words, just because a firm is an innovator in product development does not automatically make it an innovator in process technologies (in contrast to the GE example). For example, RCA, an innovative pioneer in color television and, before that, a large number of consumer products, did not adopt any of the new process technologies such as solid-state circuitry and highly automated production techniques. Nor was it the only American producer of TV sets that failed to adopt these process innovations. Thus, the United States lost this industrial sector to France (Thompson, which bought GE and RCA), the Netherlands (Phillips, which bought Magnavox), Japan (Mitsubishi, which bought Motorola), and South Korea (LG Electronics, which bought Zenith).[54] Again, the obvious historical lesson is the need to pursue both radical product and process innovations.

More generally, a study of the survival of some one hundred plants across a number of industrial sectors in the United States during the period 1972–87 found that the movement toward automation and flexible manufacturing increased plant survival.[55] But equally important was investment in RDT. *In other words, it is the combination of product and process innovation with research that protects employment the most in a representative sample of firms in the American economy.* But for the adoption of flexible manufacturing to work, a point I discuss later, also requires that companies upgrade the skill set of their workers and hire people with new kinds of occupational training, illustrating how the evolution of what is known requires continual adjustments and changes in many aspects of the industry.

The development of new products that can increase the rate of throughput, or productivity, is a particularly important way of protecting American jobs. Another American success story is Corning Glass that slowly developed optic fibers, first for the transmission of phone calls but now for the Internet and HDTV. This product innovation has changed the dominant design not only in the telephone industry but also in cable television, the Internet, and implicitly the future of the DVD rental industry and probably next the movie industry.

Unfortunately, the same success was not repeated when American companies developed the new kinds of robots for manufacturing. Since the United States invented the robots, we can of course claim credit for being innovative, but if we are unable to implement them correctly and keep on improving their design, then there is little employment gain. This story is repeated over and over again (see Chapter 6). Robots could provide considerable improvement in productivity across many of the low- and medium-industrial sectors in which so much employment was lost. Despite the development of many patents in the United States, we now import most of our robots from Japan. This illustrates again, as did the footwear example, how the means were available to protect jobs, but the CEOs failed to adopt innovative manufacturing processes. Understanding why becomes an important part of the analysis and provides lessons for the future.

In many instances, those companies that did adopt flexible manufacturing made three errors when they implemented this new process technology: (1) American engineers did not build in a large enough flexibility, thus defeating one of its major objectives,[56] (2) American managers frequently tried to use the machines as control devices,[57] and (3) the companies that designed the plant layouts for the use of these machines failed to consider the worker as part of the manufacturing system.[58] A comparative study of flexible manufacturing systems in Japan and the United States documents these errors.[59] In the American case, not only was flexibility not part of the design, but in many cases, flexible manufacturing *reduced* productivity. The American engineers that designed the flexible manufacturing systems used the same logic as the engineers that did not buy the new flexible manufacturing equipment in the footwear industry. Productivity for them meant long production runs, and therefore they did not design a system to have flexibility. This is the complete opposite of what is needed in markets that are continually and rapidly changing because of shifts in tastes and preferences (as is the case in footwear). In multiplant companies, the same production system was used in each plant and designed in central headquarters. They used the principle of "one size fits all." One dramatic example occurred in the GM plants. To correct for the poor design, one manager asked his workers to tear down the system and rebuild it so that it would increase productivity, which they did successfully. In contrast, the Japanese engineers worked closely with the plant managers and tailor-made their systems within each plant of multiplant companies. They had a very different way of thinking about what flexible manufacturing should be and how to design it. They built their systems to handle many more different kinds of parts and therefore be more flexible. Again, this failure in thinking is probably due to the way engineers were taught to think about productivity when they were being trained: "one size fits all."[60]

Another reason why flexible manufacturing did not have the success it could have had in the United States was the attitudes of the managers of the plants. Rather than allow the workers using the machines to function as skilled crafts-men, which would result in the full potential of these machines being realized, they attempted to use the machines to control the workers' pace of work.[61] To be successful, flexible manufacturing requires flexible workers with the right train-ing but, most critically, flexible managers as well. Indeed, one might argue that the real problem has been a lack of flexibility in both the managers and the en-gineers that designed the system. This is an example of how pursuing the same routines from the past, managerial control practices of workers, obstructs the adoption of radical process innovations.

Using the comparative lens again, one finds many examples of how flexible manufacturing in Germany has protected employment and wage levels against foreign competition in contrast to Britain and France. A series of comparative plant studies and industrial sector studies found that the Germans focused on the high-quality, high-price end of one market after another.[62] To do this ef-fectively, they invested in a technical vocational system that makes their work-ers quite flexible and capable of reprogramming machines.[63] While this has not made Germany completely immune to global competition, it has helped them protect their industries much more effectively than either the British, the French, or the Americans in one industrial sector after another. Again, as with Italy, the story is more complicated than just the adoption of flexible manufacturing and having well-trained workers capable of handling these new machines. The presence of associations, the close connection between the labor unions and the management of the large corporations, and a technical/vocational education system that has been in place for a century also contrib-uted to the German success. Both of these countries illustrate the importance of public-private cooperation in manufacturing.[64]

Finally, the American companies that developed these flexible manufac-turing systems did not include the skills and activities of the workers in their design of the plant layouts. In other words, not only was flexibility not part of the design process but operational engineers ignored completely the role of the workers who operate the machines. Obviously, these two errors reinforced themselves.

The historical lessons that one draws from this discussion is that it was not only the adoption of new process technologies that was important to protect American employment but also how the implementation and management of the new technology impacted on the performances that were achieved, namely, little flexibility and even a loss of productivity because implementation ignored the worker who operated the machines. To understand better why the important element of the worker was missed, we explore the dominant business model of the last four decades and its emphasis on the performances of productivity, stock prices, and rewarding individuals who succeeded with these performances.

To blame managers and engineers for these various problems does not identify the root causes and their origins. Behind the decisions that executives made is a set of performance measures and rewards systems that make one or another kind of decision logical, a topic that is discussed at length in the Epilogue. To really understand the basic problem that has resulted in the export of jobs and a declining standard of living, we must ask what was the business model for success. At various points we have touched upon some of the characteristics of this business model such as "one size fits all" and long production runs are better for creating productivity. Now I want to explain its basic logic.

The space constraints of this Introduction do not allow us to explain the origins of the general business model of cutting costs or to explicate the neoclassical theory of competition that supports it. The general business model that became dominant from the 1960s through the 1990s emphasized the importance of competing via low prices, which could be obtained by cutting costs. The success of Wal-Mart in the low-cost retail service sector is an example of this line of reasoning. It has been able to grow by constantly besting its competition on the basis of price. The example also illustrates that this business model can be appropriate in certain industrial and service sectors. But it does not work in all sectors, and as is evident in our discussion of flexible manufacturing, there is more than one way for reducing costs, particularly in manufacturing. The business model behind flexible manufacturing argues that a firm competes more effectively by quickly adjusting to new products and shifting tastes as long as the manufacturing system can handle variety.

In the Epilogue, I propose a new socioeconomic model, one based on innovation. The widespread influence of the cost-cutting model occurred because of two processes that unfolded over the four decades from the 1960s to the present. First, decision making in many organizations became captured by particular occupations, and they were able to impose the performances over which they had control as the basis of the reward system, thus benefiting their salaries as well as their power position. The first occupation to rise to power in the 1960s was the accounting profession, who were later replaced by the finance experts. The former occupation placed emphasis on profits being achieved by cutting costs, and the second substituted the idea of stock prices. Second, in the business schools, emphasis was placed on these measures of success rather than on the importance of innovation. And even in the engineering schools, the concept of productivity became interpreted as long production runs with the same product rather than being able to adjust quickly to changing market demands.

A well-documented example of the shift in power from engineers to accountants is the story of GM, which in the 1920s through the 1940s was a pioneer in product innovation. What has not been appreciated in the discussions of this company's history is that a team, a dynamic duo, led it: Alfred Sloan, the business manager who created the divisional model, and Charles Kettering, the engineer, who pursued a policy of continuous innovation. When Frederic Donner, an accountant, became CEO in 1958, product and even process innovation largely stopped in this company.[65] Unfortunately, one still observes this same mind-set in the complaints of the American car managers who resist having to develop more fuel-efficient cars by arguing that it would cost too much or is not feasible. For example, at Ford, they blocked in 2005 the development of a green car. But in point of fact, their Japanese and European competitors have already achieved many of these fuel-efficiency standards. Is it any wonder that the American automobile industry nearly disappeared? To add insult to injury, in the 1980s when finance majors became more powerful and stock prices were the measure of success, GM bought and sold over one hundred companies whereas Toyota only bought four. Clearly, focusing on stock prices and the process of acquisitions and mergers rather than on developing new cars that are successful in the marketplace was a deflection from what is important. Because this industry, which can be added to the other eleven high-tech sectors, has become a central focus in the current discussions about the economy, in Chapter 5 there is an extended comparison of how GM managed its supply chain in contrast to Toyota. And in this contrast, we see the power of networks for learning and innovation and the failure of a reward system that places too much emphasis on stock prices.

The best test for determining what kind of business model is being employed in a country is to examine how business leaders handle an economic crisis. American businesses have responded to a drop in sales by downsizing, whereas Japanese businesses have traditionally responded to the same problem by introducing new products. We have just had a dramatic example of this in the last few years when millions of workers and managers were let go. Besides laying off workers, American managers improved productivity via a variety of methods such as routinization of work tasks via the assembly line (sometimes called Fordism), increases in the speed of work, reduction in wages via deskilling, the elimination of health benefits, downsizing, the use of illegal immigrants, and so on. These policies explain why flexible manufacturing was largely resisted and poorly implemented. Over time, this model did lead to large economies of scale—provided that competitors did not create new products that were superior, which as we all know was frequently the case. Product markets frequently become more differentiated with distinctive niches and when tastes change, high sales volume is more achievable via product innovation and flexible manufacturing. Although the Japanese economy has it own difficulties and is changing as well, the Japanese business model of spurring innovation during downturns has led to the development of a number of new world corporate leaders such as Toyota, Sony, Panasonic, and Fuji, which became successful with a constant stream of innovative products, including radical innovations, with Toyota's hybrid car, the Prius, being just one recent example.

Until recently, the business curricula in the American schools emphasized productivity rather than flexibility and ignored the advantages of product innovation. This overemphasis on returns on investment also encouraged the rise of the hedge funds and the various financial instruments that produced the current financial crisis. And because of the high rewards, the best and the brightest master of business administration (MBA) students chose finance as their major. It should be observed that not all innovations are necessarily good. The innovations of various kinds of derivatives have raised serious questions. Another critical point is that these innovations were created without regulation. Innovative products need networks to ensure control and regulation.

Recently, the dean of the Harvard Business School began to question whether business education had failed because of its heavy emphasis on maximizing shareholder value and a limited understanding of the responsibility of business leaders for ethical behavior and the social consequences of business decisions.[66] He has called for a new professionalism. But this analysis fails to recognize that the real cause of the overemphasis is certain performance measures, the pursuit of productivity as the means for creating profits, and corporate actions designed to raise stock prices even if there is no real increase in value. These performances have to be supplemented and refined with radical product and process innovation and a reward system that encourages creativity.

Some readers may be thinking that if they have a college education their standard of living will then be protected and it is primarily blue-collar workers that are experiencing the consequences of the lack of innovation. The pattern of layoffs, however, in the last few years should dissuade them from this idea. But there are more ominous signs that the American standard of living is declining even for the educated. Between 1997 and 2007, *before the recession began*, the median wages for holders of bachelor's degrees, ages 25 through 34, adjusted for inflation, declined by almost 3 percent, that is, during relatively boom years, except for the decline in dot-com start-ups. One can only imagine how much wage erosion has occurred in the last three years.[67] If this were not enough, many of the high-paying research jobs are being shifted overseas. Between 1999 and 2007, foreign R&D funded by U.S. manufacturing firms grew 191 percent while their funding of domestic R&D grew only 61 percent.[68] Obviously, if this trend continues, not only will our foreign competitors reduce their unemployment, but the creation of new kinds of Silicon Valleys where learning across sectors can occur will be located outside of this country.[69]

As this brief review indicates, the reasons for the decline in the American economy are many and involve a number of factors. But the dominant theme is many missed opportunities for protecting the American standard of living. These opportunities were missed because of the maintenance of the same routines rather than changes in strategies. This "path dependency" occurred because of an overemphasis on productivity and stock prices as performances rather than on innovation and growth via creativity. But changing the performances and the reward systems is not enough. A number of obstacles have to be overcome if we are to restore the innovative edge, our next topic.

RESTORING THE INNOVATIVE EDGE:
EIGHT OBSTACLES TO OVERCOME

The narrative about the innovation crisis is a relatively simple one even if its solution is not. On the one hand, there have been a number of major structural changes produced by the six patterns of evolution described in Chapter 1. On the other hand, and for a variety of reasons, only some of which are explored in this book, individuals, managers of organizations, and policy makers continue with the routines with which they are familiar, the pathways that they have

Table I.2. Obstacles and managerial remedies

Obstacles	Managerial and Policy Remedies	Location
1. Static strategies	Seize strategic opportunities to fit evolutionary patterns	Chapter 1
2. Low-risk research	Broaden the vision of the research team(s)	Chapter 2
3. Lack of learning	Stimulate cross-fertilization of ideas	Chapter 3
4. Stovepipes	Integrate the organization	Chapter 4
5. Reactive leaders	Appoint transformational leadership teams	Chapter 4
6. Valley of death	Construct appropriate coordination practices	Chapter 5
7. Lack of industrial policy	Cooperation between the public and the private sectors	Chapter 6
8. Nonvisible blockages	Perform timely feedback on organizational blockages	Chapter 7

been taught, and thus do not adapt to these changes. We have to break out of these routines, and determining which routines have to be changed depends on understanding the obstacles to innovation.

Earlier I suggested that one of the major definitions of management is solving problems. In the case of restoring the innovative edge, it means eliminating the many obstacles that are preventing innovation. Table I.2 lists eight sets of what, I think, are major obstacles that reduce our capacity to remain an innovative nation. As I said earlier, the objective of this book is to suggest remedies to restore the innovative edge to the United States to prevent further declines in the American standard of living. Each chapter deals with one or more of these obstacles.

A good starting point in the discussion of obstacles, the *first* one, is the static organizational strategies and federal and state government policies. It is not that they do not change, but frequently they fail to change in the right direction, which is in line with evolutionary patterns that exist. Chapter 1 outlines six evolutionary patterns: changes in consumer tastes or demands, changes in the number of products available to satisfy these demands, the globalization of RDT, the differentiation of the idea innovation network, proliferation of special noneconomic concerns, and new stages of manufacturing. Creating new strategies to speak to these evolutionary patterns allows managers and policy makers to get ahead of the innovation curve. In other words, these competitive difficulties also represent opportunities when properly understood. They can provide comparative advantage.

The *second* obstacle shifts from the societal level of static strategies to the level of the research team. The problem at this level is more the reluctance to take risks and the emphasis on what is called "normal science," incremental advances or minor product modifications. The formula for developing more radical innovations or particular kinds of scientific breakthroughs in teams would appear to be simple—broaden the diversity of the team's perspectives—but it presents a number of decisions and dilemmas. Chapter 2 provides a checklist for managers to consider that describes the different ways in which diversity can be achieved. Adding diversity can breakdown the rigid models of thinking associated with each scientific discipline or professional/managerial occupation.

Adding diversity must be done relative to the nature of the research problem and the team that is required. One size of research team does not fit all projects.[70] In addition, to the degree of radicalness desired, two other dimensions provide managers with a way of classifying the nature of their research teams: the scope of the problem and whether the team should cross organizational boundaries. Various dilemmas are involved in these choices, and these vary by the kind of research organization.

The *third* obstacle is the level of interdisciplinary learning that exists within the research teams. In one sense, this problem is created by adopting the solution to the previous obstacle, increasing its diversity. As diversity increases and the vision is broadened, it also tends to dampen communication, and with less communication, less learning, and of course, less radical innovation.[71] As diversity increases arithmetically, communication difficulties grow geometrically. Thus many research teams avoid this problem by opting for incremental advances in either science or in technology; it reduces communication problems and makes it easier for each researcher to work on his or her own.

The previous two obstacles are located at the research team while the next two shift to the organizational context in which the research teams are located. The first obstacle at this level is the tendency for silos, or "stovepipes," to develop, that is, for research teams and especially departments and divisions to act largely independently of others in the same organization. Various managerial practices for integrating across research divisions are illustrated in a detailed case study in Chapter 4. This case also illustrates both successes and failures and thus indicates additional kinds of blockages that exist.

An even bigger obstacle, and one common to many executive directors, whether in the public or the private sector, is the tendency to be reactive to changes in the environment rather than proactive in anticipating and solving problems in the larger societal context. Instead of adapting to the environment, transformational teams are needed to change the society and solve global problems. This obstacle is also discussed in Chapter 4.

The next two obstacles are at the meso or sector level. One of these obstacles is the development of communication gaps between research arenas as they become differentiated into separate organizations, which has been called the *valley of death*. If organizations represent groupings of a number of research teams, then within the idea innovation network at the sector level we need to create coordination models as a way of stimulating cross-fertilizations, not just within and between teams but along the entire network. Because failed evolution is such an issue in the United States, Chapter 5 discusses examples of both failed evolution and successful evolution, namely, when leaders of organizations changed their strategies so that they could continue to produce radical product and process innovations. In this chapter, the major theme is the steady movement toward more complex problems, as outlined in Chapter 1, one that necessitates more complex arrangements to coordinate the idea innovation network.

In Chapter 5, a new topic is added to that of the idea innovation networks, supply chains. In complex products, the supply chain has to become a learning chain because this is the heart of the matter of retaining good jobs in the United States. Foreign firms have been taking over the production of the less sophisticated components and then moving up the supply chain to the more technically advanced products until much of the entire product is produced outside the United States.[72] This leads naturally into the discussion of the next obstacle.

A second key obstacle at the meso level, the topic for Chapter 6, is the relative lack of enough public-private cooperation in developing radical innovations. If the United States is to protect employment and if it is to export, it must restore its manufacturing capacity, which requires public-private cooperation because neither sector can accomplish this societal goal alone. Both Germany and Italy provide different models of how the public and the private sectors cooperate. The example of Taiwan is given in Chapter 5. Chapter 6 discusses four policies in which more public-private cooperation could and should be generated: (1) create joint public and private sector coordination of the idea innovation networks attached to sophisticated components in the supply chain, (2) emphasize more manufacturing and quality research in various cooperative agreements between the national research laboratories and firms and in other programs such as the Small Business Innovation Research and Advanced Technology programs, (3) create extension services for those industrial sectors that have many small- and medium-sized companies, and (4) collect data on the evolution of the idea innovation network.

The eighth and last obstacle is the failure to recognize organizational blockages quickly. The remedy for this is in-depth evaluations to detect these blockages. The recommended new data at the sector level to be collected by NSF would provide a framework for detecting lack of technical progress in specific arenas but not necessarily the reasons. These in-depth studies should be conducted frequently. One obvious reason for timely and frequent feedback is that scientific discoveries and technological breakthroughs are constantly reshaping the competitive marketplace. As these emerge, the managers of research organizations and firms must reevaluate their choices of strategies, that is, avoid the tendency to keep the same routines. Not all of them may need to be changed, but at least they should be reconsidered. The objective is to find the causes of failed evolution. In each of the chapters, the reasons for obstacles are suggested, indicating how complex the problem of restoring the innovative edge is.

The metaphor for overcoming the eight obstacles is solving the puzzle of the Rubik's Cube. These obstacles are themselves interrelated in complex ways. Overcoming one creates new ones. Remedies have to be revisited precisely because the four levels are interrelated in a number of ways, some of which are still not well understood. But together these eight obstacles provide a start for restoring the innovative edge. The objective of this book is to begin the process of overcoming obstacles and to think about scientific and technological policy in new ways, laying a foundation for the new *science* of science and innovation policy, the concluding topic.

A NEW POLICY PARADIGM TO PROTECT AMERICA'S STANDARD OF LIVING

The advantages of having viable high-tech innovation sectors in our innovation system are obvious. High-tech industries are associated with high value added because the companies that launch radically new products first can charge higher prices and they also tend to export more because there is likely to be global demand for innovative products on the cutting edge. Indeed, increased exports would be the major indicator that we have restored our innovative edge, a movement of our balance of payments into a positive territory. Then, too, high-tech occupations pay higher salaries and allow for a middle-class lifestyle. Finally, and as others have argued, the learning associated with innovation tends to "spill over" into other industries, affecting their innovativeness and productivity as well.[73] This is one reason the current exporting of R&D jobs overseas is so detrimental to the future prosperity of the country. And it is for this same reason that I discuss a remedy, establishing more Silicon Valleys, in Chapter 6.

To protect the American standard of living, policy makers need new models, predictive models, so that they can take corrective action more quickly, preventing problems before they become crises that require much more massive interventions. For example, policy makers during the middle of the decade began to recognize that it had an innovation crisis, one that became abundantly clear as the trade balances across all high-tech sectors turned negative. If the first signs of this trend had been recognized and diagnosed with a predictive model in the 1990s, perhaps policy interventions might have diminished the extent of the current crisis, one now obscured by a number of other economic crises and political problems. Consistent with this need for predictive models to anticipate problems in science and technology, John Marburger, the former head of the Office of Science and Technology in the United States, has called for a science of science and innovation policy. In response, the NSF has created a program of research to build such a theory. Interestingly enough, at about the same time, a spate of articles appeared that advocated theory-led evaluations that captured the complexities of the innovation processes.[74] In particular, one evaluator called for identifying the blocks and obstacles that required policy correction.[75] All of these trends are themselves propelled by the increasing emphasis on accountability for investments in science and technology, not just in the United States with the Government Performance Results Act (GPRA) of 1993 system but also in the European Union.

What is especially challenging about developing a *science* of science and innovation policy is that scientific and technological systems themselves are continuously changing. And although the investments in science do not always translate into new knowledge, they have been expanding more and more rapidly in the last few decades (see Chapter 1). As the frequency of radical innovations—product or service and process or provision—increases, product development times have been declining, and the product life cycle is being compressed. How does one build a model, a predictive model, about a rapidly changing system? My answer is a Darwinian one: The predictive theory must be an evolutionary one, in this instance the evolution in how scientific breakthroughs and radical technological advances are produced. To take corrective action, policy makers need to recognize where evolution is happening (i.e., which sectors and, more critically, where it is failing to evolve, not only the sectors but the obstacles and blockages that explain failure and which need to be corrected).

However, it is not science and technology writ large that evolves but rather specific scientific disciplines and technological complexes, and therefore policy makers should shift their focus from the usual one on the national system of innovation to the specific sector and beyond this to the noneconomic sectors involved in national problems or global concerns.[76] This becomes abundantly clear when one examines the trade balances in specific high-tech sectors in Europe and the United States; semiconductors, pharmaceuticals, defense, telecommunications, Internet services, automobiles, and aircraft vary in the investments in science and speed of product development as well as in their trade balances, as reported in Figure I.2. Therefore, the predictive model should be at the meso level where the largest variation in the pace of development in science and technology occurs. At this level, one can more easily identify the blockages and obstacles that prevent evolution, although admittedly some of these may adhere in blockages at the societal level. In other words, to return to the innovation crisis in the United States, one could have focused on the high-tech sectors where the negative trade balance first occurred.

For policy makers, however, the key insight of this model is the building of an argument for the need for a new policy perspective, namely, the twin concepts of evolution and failed evolution in how radical innovations are produced. The continuous changes implied by the six patterns of evolution presented in Chapter 1 and in particular the evidence for how the production of radical innovations alters, including the contrast between different kinds of high-tech sectors, indicate the advantages of adopting a meso or sector perspective in policy analysis. The description of these changes allows policy makers to identify where evolution has failed. Because most individuals, managers, and policy makers continue with their same routines, their inertia creates obstacles that have to be remedied for progress. Thus, the concept of failed evolution focuses attention on this tendency, providing policy makers with ways of thinking about how to overcome these obstacles.

But it is not only the evolution of how innovation is produced that needs to be added to the policy model but also the recognition that there are there more arenas of research besides the usual focus on RDT, that is, basic science, applied research, and product development. This model adds the arenas of manufacturing research, quality research, and commercialization research or the need to collect data on BAPMQC, as I have indicated. This helps identify lacunae in our current policy efforts, especially in industrial policy. In Chapter 6, a number of policy recommendations are made about how to change our industrial policy in line with the implications of the evolution in science and technology.

The eight remedies represent a much more coherent policy perspective for restoring the innovative edge than what is usually proposed in the policy literature. And since each of the eight obstacles is caused by multiple blockages, many of which are discussed, policy makers have a good framework for analyzing failed evolution. But these eight sets of obstacles and blockages are hardly all that exist. In particular, to begin to build a more systematic sense of the kinds of blockages that they are, especially at the team and organizational level, Chapter 7 discusses how a number of these might be measured.

Attached to this new policy model is not only a set of recommendations about changes in our industrial policy but also a new policy evaluation model that places an emphasis on timely feedback on the measurement of technical progress. Here is where there is a need for a lot of new policy research, attempting to develop metric systems for measuring technical progress in real time by other than papers, patents, citations, prizes, and the usual criteria. Some examples of how this can be done are provided in Chapter 7, but it is only a beginning.

Together the concepts of evolution and failed evolution, the listing of eight obstacles and their alternative remedies, recognize that the innovation process is a complex one and that there are multiple reasons why the United States is falling behind. It is not just a question of lack of expenditures or the lack of interest of college students in science, although these are factors. This is one reason that, in this Introduction, I have reviewed the explanation for why the United States lost many low-tech jobs to countries overseas. It is to highlight the very important lesson that a number of decisions about innovation are made in the various research organizations and business firms in the United States and have little to do with federal government policy, and to indicate that alternatives existed that would have saved at least a certain proportion of the manufacturing employment in this country.

Behind the eight remedies is a new logic or social science paradigm that focuses on innovation or the resolution of problems rather than productivity or the conservation of resources. A slight change in the angle of vision leads to quite different ideas. Various pieces of this new logic is a leitmotiv throughout the many chapters, frequently in the conclusion to each chapter indicating how this new perspective speaks to various issues in a number of social science literatures. Placing these discussions at the end allows those who are not interested to skip over them. But in reading them, one learns how this book is designed to speak to a number of different audiences. Beyond this, the Epilogue pulls together many of these ideas into a new paradigm of socioeconomic change built upon the importance of innovation for economic growth.

1 ADOPTING STRATEGIC OPPORTUNITIES THAT FIT EVOLUTIONARY PATTERNS

Step One in Restoring the Innovative Edge

ONE OF THE greatest challenges for academics is to develop an evolutionary theory about how science and technology are evolving, the topic of this chapter. Perhaps the biggest obstacle to restoring the United States' innovative edge is the tendency for business managers and policy makers to maintain the current strategies despite the many societal and global changes around them. If they change strategies, usually little consideration is given to how the world is changing and, more particularly, in what direction. Paradoxically, this lack of perspective is especially true of the many books and reports on the innovation crisis, which one would assume would have a view of evolutionary patterns. Most of them do report the growth in RDT expenditures and the globalization of competition, but in some respects these are not the most important evolutionary patterns for the purposes of shaping business strategies and federal and state government policies. All that these patterns do is indicate the spread of competition about innovation throughout the world.

Being successful and getting ahead of the innovation curve means having some information about several other evolutionary patterns. But this requirement raises a real doubt: Is it possible to predict the future? The answer is a qualified yes, meaning within general guidelines about evolutionary directions. I do not believe very much in what might be called technological forecasting that makes specific predictions that we will develop fusion energy or a cure for cancer or a substitute for copper within a certain time period. But I do believe that one can observe evolutionary patterns, and, if they can be explained, one can then select strategies that complement these patterns. These allow CEOs and policy makers to anticipate future directions of changes in tastes and also likely obstacles in being able to produce radical innovations.

The following six evolutionary patterns, or trends, are observable:

1. Growth in global research expenditures and therefore in the number of radical innovations
2. Globalization in the development of radical innovations
3. Differentiation in consumer tastes
4. Changes in the way in which radical innovations are produced
5. Expansion in the number of noneconomic concerns
6. Emergence of a new stage of manufacturing

The pace of these changes varies in each of the sectors, both high-tech and others, as we can see by simply looking at the differences in the amount of money invested in RDT within each sector (see NSF.gov/statistics).

Recognizing these evolutionary patterns provides some additional information for policy makers in the debate about "picking winners." Just as technological forecasting is suspect until proven wrong, federal government RDT investments in specific firms (except in those few industrial sectors where there are only one or two firms; see Chapter 6) could indeed be risky. But recognizing evolutionary patterns and facilitating all the firms in a particular sector avoids the pitfall of picking specific winners. And while it may seem that investing in particular sectors is another variation on the theme of "picking winners," it is not because the high-tech sectors are clearly visible globally. Or to put it in other terms, most countries appear to have the same list of high-tech sectors where RDT expenditures are growing, with nanoscience and nanotechnology at the top of this list. More fundamentally, government support of these evolutionary patterns would allow American firms to compete on a more even playing field with the European and Asian firms (see Chapter 6).

The first two patterns, the explosive growth in RDT expenditures and radical innovations and their globalization, mean that not only many more radical product and process innovations are launched on an annual basis but that increasingly they are coming from overseas. The rates of growth are uneven across countries and especially within sectors, but the decade averages in real terms and as a proportion of the nation's GDP are relentlessly increasing. Between 1984 and 2004, the expenditures of the United States almost doubled, going from $152 billion in constant dollars to $286 billion, while Japan's did

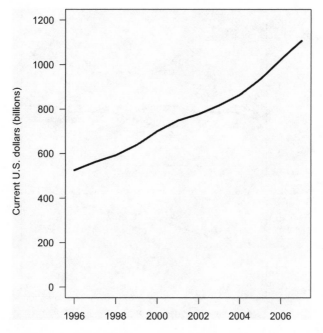

Figure 1.1. Global expenditures in RDT
Source: Generated from NSF 2010 data.

double, moving from $53 billion to $105 billion.[1] Just taking the last eleven years, the global expenditure has doubled (see Figure 1.1). At the same time, the number of countries involved in research is also expanding to include South Korea, Brazil, New Zealand, and of course China and India (see Figure 1.2).

Perhaps more revealing is the fundamental alternation in the relative importance of RDT investments as measured by the percentage of GDP allocated to research (see Figure 1.2). The percentage of GDP allocated to RDT expenditures over time has remained largely stable in the European Union and the United States despite the desires to raise it to 3 percent.[2] For the United States, the percentage is 2.62, and for the Organisation for Economic Co-operation and Development (OECD), it is 2.25.[3] The European Union set 3 percent as a goal in the Lisbon Accords, and recently the Obama administration also chose this figure as a goal. But given the current economic crisis, it is unlikely the United States or the European Union will reach these goals by the dates that they have set. Thus the growth in other countries becomes more ominous for the competitive position of the United States. As can be seen, several of the Asian countries,

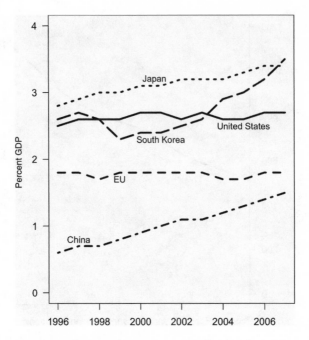

Figure 1.2. Percentage of GDP spent on RDT for selected countries and regions
Source: Generated from NSF 2010 data.

most notably Japan, South Korea, and China, have steadily increased their expenditures as a percentage of their GDP. Japan is now at 3.3 percent, Korea is at 2.96 percent, and Taiwan is at 2.46 percent. Not shown within the European Union, the figures for Sweden (3.89 percent) and Finland (3.48 percent) are also relatively high, and one might add that these countries are doing comparatively well in international competition. The highest expenditure is in one of the smallest countries, Israel, at 4.49 percent, and that is probably an underestimate. Together these two evolutionary trends mean that not only is the amount of investment in research in the high-tech sectors rising, but research is becoming more relevant even in the medium-tech sectors. One of the more important implications of this globalization in research is that, increasingly, U.S. firms will have to form global networks to remain competitive, which, as we have seen, is one reason why they are investing in RDT overseas.

The first three evolutionary patterns given on the list represent ways in which the nature of market competition is changing. Price as the determining factor becomes less and less important, especially in those high-tech sectors that are our greatest concern. More and more, being first with the most sophisticated technology—provided there is product quality—gains more market share, trumping product price as the determining factor. Thus an economic growth strategy based on developing new products rather than a productivity strategy of cutting costs is the surer road not only to profits but to growth in employment as well. But being first is not enough; manufacturers and service providers must *continue* to advance technologically. The failure of static strategies is illustrated by a study that showed that each dominant disk drive company failed to move to the next smaller size, that is, from 14-inch to 8-inch, then 5.25-inch, and on to 3.5-inch, and after this 2.5-inch, leading to their loss of market share. While admittedly each new advance meant a disruptive technology, the shifts should have been predicted on the basis of the needs of different segments of the market and the constant demand for smaller size, greater simplicity, and technological advance.[4]

In fact, a positive example of how to do this is provided by Hewlett-Packard (HP) and its solving difficult problems in producing printers for disparate users. Various models of their printers can print billboard advertisements, textbooks, or labels for Heineken beer bottles.[5] Next, customized wallpaper for children printed by HP printers will be offered in the children's chain stores of Nickelodeon and Dr. Seuss Enterprises. Customers can go to a website and enter the dimensions of the room along with a selection of the children's favorite characters and then pick up the customized wallpaper at a local store for their children's room. This example illustrates the importance of developing new kinds of distribution channels for radically new products as well as the principle of customization having superior market value.

The strategy of recognizing different customers has also been pursued by Nokia and its constant technological advance in cell phones, largely passing Motorola, which has lost market share in this very important and growing market niche. In turn, Apple with its iPhone has reasserted American leadership in design and versatility in this critical market. These examples also illustrate how, with global competition, managers need to plan on sets of products designed to meet both different cultural tastes and disparate national technological requirements. Cell phones are a perfect example of the increasing need for products and services that are designed to operate with varied technological platforms as well as cultural tastes.

Although the growth in RDT expenditures and its globalization has been widely recognized, the third pattern has been less often discussed and is therefore one of the foci of this chapter. By carefully explaining why and how consumer tastes are evolving toward more demands for technological sophistication, product and service quality, good design, and above all personalized products and services, managers are in a better position to plan their research to meet these demands and gain comparative advantage.

Likewise, understanding the consequences of continued growth in expenditures on RDT for how radical innovation is produced both within countries and globally suggests how this planning can be implemented, that is, what kinds of research teams and networks need to be constructed to take advantages of new ideas within one or more of the six research arenas indicated in Table I.1. The construction of bridges via networks to overcome the developing interdisciplinary communication gaps created by this evolution presents one of the most important tasks for managers of research organizations and firms. It is the failure to evolve to meet the new circumstances created by evolution that explains much about the innovation crisis in the United States as well as elsewhere.

The evolution in tastes and the way in which radical innovations are produced are not the only evolutionary processes unfolding. Some of the more interesting trends involve the growing concerns about the impact of products on the environment and on health. These impacts are created while producing the product as well in its operation, and they are hidden costs that are not part of the price charged for the product.[6] The most dramatic of these hidden costs are various forms of pollution (from trash on the ground to smog in the sky), consequences of global warming (from melting ice caps at the poles to cyclones and hurricanes in the middle of the planet), dependency upon oil (from engaging in destructive wars in the Middle East to oil drilling in Alaska), the use of nonrenewable resources (from wood to metals), and health risks of various kinds (toys with lead paint to Love Canal) for the consumers. The recent oil spill in the Gulf of Mexico has educated everyone on the importance of hidden costs. Some of the biggest opportunities for restoring the innovative edge are in developing new manufacturing techniques that reduce risks to the individual and the environment. And, clearly, solving the problem of how to prevent a repeat of what happened in the Gulf of Mexico will require a number of scientific breakthroughs to learn how to cap wells 5,000 feet under the sea.

In addition to growing concerns about the hidden costs in the manufacturing of products, a number of global problems have emerged. The beginning of the twenty-first century saw the emergence, or perhaps better understanding, of some new problems, including terrorism, tribal wars, and pandemics such as AIDS and the avian flu. The most striking aspect of many of these issues is that they are *truly global problems*, that is, not one country, not even the United States, can solve them unilaterally. Cooperation, as well as scientific breakthroughs and technological advances, then becomes a key to making progress in solving these different problems.

Although I later suggest that being first with technological sophistication, good design, and other product factors such as reduction in carbon footprints provides protection against price being the determining factor, especially in the high-tech sectors, we must also recognize that price remains important, particularly in the low-tech sectors where most jobs are being exported. How does one maintain productivity and yet produce different sets of products for different parts of the world? The answer remains radical process innovations such as flexible manufacturing. But because the United States largely failed with the implementation of this area, as discussed in the Introduction, we seize the initiative and attempt to develop the third industrial divide, which involves being able to change the properties of parts on the assembly line.

We have, then, four strategic imperatives for restoring the innovative edge, each of which flows from a specific evolutionary pattern:

- Customize products and services.
- Recognize the evolution in the patterns of doing research.
- Solve national concerns and global problems.
- Increase flexibility in manufacturing by being able to change the composition of parts.

The evolutionary forces behind these imperatives are explained in the following four sections of this chapter.

THE EVOLUTION OF TASTES

In the Introduction, two major reasons were cited for the loss of American employment during the 1960s through the 1990s: failures to adopt radical product and especially process innovations and the consequences of the emphasis on cutting costs and increasing stock prices. Still a third reason for the loss of many jobs was the strategic failure to recognize the *direction* in the evolution of consumer tastes, and a fourth was the failure to adapt products to the international tastes of other countries. The failure of American car manufacturers to make cars with the steering wheel on the right side for Japan is well known. But the more critical failure is not recognizing the movement toward consumer demand for new technological advances, higher quality, good design, and, most important, individualized products and services. The continued purchases of the latest computer, video game consoles, GPS, cell phone, and high-definition television are illustrations of the importance of technological advance. Frequently, American companies did not shift their strategies toward higher quality until it was too late. Germany was able to protect its much higher wages by positioning itself in the high-quality end with individualized products in such areas as kitchen design, clothing, and, above all, machine tools.[7] The continued penetration of the American automobile market by German and Japanese manufacturers occurs in large part because of their innovations, including those in fuel economy and the quality of their cars and their design.

The causes of the evolution in consumer tastes in these four directions can be explained by the growth in education and the increased occupational specialization.[8] With education comes a more rational approach to weighing the costs and benefits of particular purchases and thus a concern about quality as well as an appreciation of technological advance. With occupational specialization comes a desire for a more specialized or individualized lifestyle. Some American companies have built their fortunes on these trends. Starbucks is only one example, and the microbreweries are another. And while fast food restaurant chains will always have a place in the food alternatives that Americans want, it is also true that gourmet restaurants are becoming more and more popular, as is the culture of wine. The success of Honest Tea and its policy of creating new kinds of drinks annually, as reported in the Introduction, is an example. All of these illustrations, from the price of Starbucks coffee to the price of a good bottle of wine, indicate that the educated consumer is willing to pay more money for an individualized experience that is perceived to be of high quality. But they also illustrate another important strategic imperative, the differentiation of tastes and the need to provide a wide range.

One of the major failures of the American companies was staying with their static strategies that emphasized a mass market with one or two basic models for the average American instead of replacing them with multiple models for different segments of the market that reflected differences in education, income, and lifestyle as well as needs of particular users. Furthermore, they did not attempt to create new products in these areas, which might have allowed for higher prices and thus have protected the wage levels in other ways. But it must be recognized that to question the strategy of a firm that has been successful in the past is difficult to do and requires a very flexible mind that is open.[9]

One of the major stumbling blocks for both businesses and government elites, as I have observed, is a strategy of "one size fits all." In one sense, this is strange given the long history of marketing research that discovered the Marlboro Man and identified that Oldsmobiles and Buicks appealed to different occupational sectors. Market researchers have long observed that tastes vary as a function of the many ways in which we can describe the differences between people: age, gender, race, religion, education, and of course income. But they have not had a theory about the *evolution* of tastes nor, in particular, about some of the most striking preferences that have emerged during the past several decades: increasing demands for individualized products and services, higher quality, the latest technological advance, and good design. Of these four trends in the preferences of the consumer, probably the most important is the increasing demand for individualized products and services.

What is the explanation for this evolution? America built its huge economic success in part, as we have seen, by emphasizing mass markets and "one size fits all." Obviously, for a period of time this was a successful strategy, which is why it is difficult to abandon it. What changed to discredit this strategy? The nature of the consumer. The United States in particular and to a lesser extent many other societies in the first few decades after the end of the Second World War were mass societies in which the wants and needs of the population for many basic commodities were similar: essentially, a cheap car, one's own home, and the basic utilities of modern life such as telephone, radio, and then television. As a consequence, these products could be and were mass produced without much variation.[10] Tastes were similar not just because of lower income levels but, more critically, because of the lack of diversity in both the American education system, where there were few technical or elite tracks at the secondary level, and in the kinds of jobs that were available after leaving high school. Whether clerical or blue-collar jobs, the work was largely routine and essentially similar. The similarity in life experiences led to similar tastes in the choice of one's lifestyle.

This apparent uniformity in tastes started changing with the rising levels of education because individuals who graduated from college were provided with many more occupational and professional choices. In other words, diversity in the way in which the consumer was educated increased, and because of this diversity in occupational and professional occupational choices, which became very important parts of one's identity, diversity of lifestyles followed as well. Just as Oldsmobiles and Buicks were associated with distinctive lifestyles, the many greater occupational choices have been reflected in a proliferation of tastes in cars, providing European and Japanese manufacturers with opportunities to pierce the American market with distinctive models. In addition, the growing leisure-time choices had a similar effect. In this area, the range of options has been steadily expanding every year with elder hostels, ski trips to various parts of the world, environmental cruises to the Amazon or the Antarctic, stock car races, safaris in Africa, cultural cruises on the Rhine or the Nile, long train trips across Australia or Russia, and so on. Nor does this trend only involve adults taking vacations; increases in the variety of leisure-time activities for children are also remarkable. The most visible signs of this variety are the many choices provided in the kinds of amusement parks, summer camps, and school trips for children. The rise of the soccer moms in sport utility vehicles (SUVs) reflects one extension of this diversity, namely, in children's sport opportunities. The much greater diversity of both work and leisure-time identities meant a movement toward individualized tastes in many other aspects of life as well. And thus the demand for personalized products and services, which is called mass customization because of the extent to which this was occurring in formerly mass-produced products that previously had little differentiation.[11] The example of HP cited at the beginning of this chapter indicates how important it is to meet the needs of a variety of consumers as well as the advantages of customization, which allows for expanding market share.

The differentiation of ever smaller sets of the population with similar tastes is not just a function of the different occupational and professional choices as well as leisure-time options. Sometimes the recognition of what is missing is simple, as in the example of Under Armour producing performance-enhancing clothes for various sports teams, as noted in the Introduction, just as previously Nike pioneered performance-enhancing shoes for runners. The process of differentiation flows not only from the variety of different kinds of leisure-time activities that need their clothing and shoes but also emerges from the growing recognition of new kinds of populations that had been previously ignored by market researchers. *Microtrends* delineates a number of these new population categories, such as the soccer moms cited previously.[12] Even the most basic social categories of age and gender are moving toward greater differentiation because of distinct needs and tastes. Because a larger number of people are living until an older age, seniors are divided into different categories based in part on their health status, such as the frail elderly. This trend has meant building new kinds of residential facilities that recognize the gradations in health that exist toward the end of one's life. At the other end of the age spectrum, the earlier and earlier age of sexual maturity has resulted in the differentiation of adolescence as a distinctive category. Yet paradoxically, many youth remain adolescents for a much longer period, which has created a trifurcation of marketing strategies into preteen, teen, and young adult. Consider gender, which we might assume is the simplest dichotomy of all, male and female, but in fact now many variations on this theme are recognized within the bisexual, gay, lesbian, and transgender (BGLT) community.

If the diversity of occupational and leisure-time choices for the college educated drives the need to create many different kinds of products and services of a customized nature, the higher educational levels rather than the higher income levels are driving the other three evolutionary changes: demands for higher quality, more sophisticated technology, and better design. Income differences have long been associated with different versions of many products and services. But in general, luxury models of cars, boats, vacations, and similar products and services are not the really interesting variations in tastes that are occurring. It is the demand for much greater quality in cheap cars, boats, and vacations that is the key. Quality has always been available in luxury brands and is taken for granted; quality in a Saturn rather than a Cadillac is a new kind of demand. The same can be said for technological advance and good design.

Many definitions of *quality* exist, especially as one moves away from the product sector to the service sector. But in the area of products, quality can be measured in at least four distinctive ways:

1. Reducing defects in the product during the manufacturing process
2. Lowering the repair costs during the use of the product after sale
3. Expanding product life
4. Lowering the operating cost during the product life

Concentrating on these last three aspects of quality can give American manufacturers a considerable edge in the marketplace. But to make improvements in these areas probably requires a considerable amount of research on how to do this, as discussed in the next section.

These same ideas can be applied to services as well. Consider the case of teaching a particular skill such as solving problems using algebra. The given list defining quality would translate quality in teaching into fewer dropouts from the course, less remedial time spent learning, longer retention time of the ideas, and more efficient use of the principles. Again, we have an area that needs a lot of basic research to develop radically new kinds of teaching techniques for quite disparate student populations—the autistic, underprivileged boys, teenage moms, and so on.

The simplest measure of technological advance is how much improvement there is in *functionality* of the product relative to its predecessor or competition. In our research at a national research laboratory, we found a number of illustrations of how to measure this improvement in functionality: in research on catalytic combustion, the number of particulates that are entrapped is the salient measure; in the microfluidic ejector project, it is smallness of the drops and their viscosity; in the work on nanoscale energetic materials, it is the number of particles that have both oxygen and explosive material.[13] In some instances, the functionality reflects the replacement of one component with another. An example is the replacement of some analog components with digital components in radar systems. Or functionality might involve the synthesis of components, such as the combination of two components in the development of semiautonomous modular robotic control technologies.

Because so many products are becoming more and more complex as additional attributes are added, multiple functions are built into the same product much on the model of the Swiss Army Knife. Consider the combination of fax machine, printer, and copy machine or the mixer that with various attachments can handle normal to heavy mixtures, whip air into mixtures, and mix and knead yeast dough, to say nothing about making ice cream, raviolis, or pasta. The development of glass fiber cables has not only created the ability to handle much heavier loads of telephone traffic but made possible the addition of television and the Internet on the same line. But perhaps the most dramatic illustration is the large number of applications that can be used on iPhones—225,080 at the time of this writing.[14]

Another source of increasing product complexity is the addition of numerous control devices that are incorporated into the product, with the extreme being the many electronic controls of the airplane. These complexities make the supply chain a critical component in creating radical process and product innovations. For this to take place, the supply chain has to become a learning network so that it can become an innovative supply chain; if it does not, the suppliers become another drag on American competitiveness. In the second section of Chapter 5, the Japanese automobile supply chain provides a model of what is possible.

One might assume that the latest technological advance comes in one attribute, but this is not the case. Sophistication in products leads to the proliferation of attributes because, again, the demand for individualized tastes manifests itself also in which technological advances are considered most attractive. A good example is the current competition between plasma and liquid crystal display (LCD) televisions that are being produced in a variety of sizes. Likewise, technological advances in laptops mean not only the increased memory of the chip, a typical way of describing advances in computers, but the varying sizes of screens; their slimness, weight, battery duration; the sophistication of the operating system; the variety of available programs that work with the software, and so on. Furthermore, these attributes are desired in different combinations. In a study that we conducted at a national laboratory on measuring technological advances across a number of different products, many of which were being developed for the government, the following generic themes were common examples of technological advance:[15]

- greater reliability
- higher efficiency
- better precision
- more flexibility
- increased miniaturization, or reduction in size
- ease of use
- speed of use

Each of these advances in technology can be applied to many products and also to services.

The importance of design has been aptly demonstrated by the success of Apple in the design of its many products from the laptop to the iPod to the iPad. This is not to say that good design is important for everyone but only to indicate that it is one of several major attributes upon which products are evaluated. Ikea has built an international reputation on the design of its products that are geared to the cost-conscious consumer. This is another example of where European manufacturers, not just China and India, have made inroads in the low-price markets in the United States.

Each of these evolutionary trends in the tastes or the kinds of demands that consumers are making should inform the way managers of firms think strategically. How can one increase customization? Let me provide one example in one of the high-tech sectors that is a focus in the next section, the biotech and pharmaceutical sector. The principle of customization suggests that in all treatment therapies one would search to see if a particular therapy, whether drug, surgical intervention, or other form, only works with specific kinds of individuals defined on the basis of either their personality (A types), or genetic predispositions (susceptibility to a specific cancer), or some other way of appropriately classifying individuals in a particular morbidity. Too often drugs are given to everyone when in fact they are likely to work with only a certain percentage of the individuals. Drugs and other therapies should be tailor-made for the individual, including designing special drugs for individuals who experience side effects. Although this clearly does not apply to every medical therapy, it does apply to many and represents a new way of thinking about medical research, one that is actually becoming more and more common as medicine recognizes that "one size does not fit all."

Beyond health services, the manufacturing of radical new products that have (1) much better quality, as defined in any of the four ways listed previously, (2) greater technological sophistication, again as previously defined in a variety of ways, and (3) much better design will gain price premiums in the market place. The more these different ways of speaking to the individualized needs of the consumer increase (while respecting cultural differences and national technological requirements), the more we can begin to restore our innovative edge.

THE EVOLUTION IN HOW INNOVATION IS PRODUCED

The biggest challenge for R&D managers, policy makers, and academics is to keep aware of the ways in which radical product and process innovations are being produced in specific sectors. As stated in the Introduction, one of the main objectives of this book is to help managers and policy makers in this task. One of the big trends that is unfolding in many, if not all, sectors is that radical advances in one arena do not move forward in a linear manner anymore but instead move backward and forward with multiple feedbacks among the research arenas that are differentiated in a particular sector, which is one of the major ways in which sectors vary, as is indicated later. Hence, the term *network* rather than *chain* becomes a better description of the process of how

Growth in knowledge leads to

- **differentiation of research organizations**
- **evolution of new networks to connect those organizations**

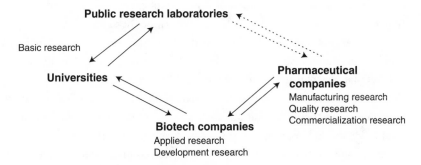

Figure 1.3. Evolution in knowledge production

radically new ideas in any one arena are translated into radical product and process innovations. But this nonlinear movement also means that coordination by markets, especially when speed is imperative, is no longer viable, and increasingly network coordination becomes the preferred method of ensuring enough interaction for rapid development of innovations.

The evolutionary pattern described in Figure 1.3 in the biotech and pharmaceutical sector is well known.[16] Scientific research on DNA and RNA in the 1950s and 1960s led to the new paradigm of molecular biology. This was followed by the creation of a new occupational specialty that spread throughout a number of the traditional areas of research in biology and afterwards invaded medical specialties as well.[17] Associated with this new paradigm were new research techniques such as genetic modification and manipulation, genetic engineering, recombinant DNA, and cloning that were relatively cheap and easy to perform.[18] As a consequence, small biotech companies were founded to do applied research and product development. Parallel with this R&D was the differentiation of a new source of capital, venture capital, which provided the initial funding. Not unexpectedly, these biotech firms tended to be located around universities that had large amounts of funding for basic biological research and where scientists could more easily establish their own firms to exploit their basic research ideas developed in the university.

Small biotech firms could develop new products, but if they were to be applied to humans or even applied in large-scale use for animals, they could not afford the cost of clinical trials nor handle what is called the "scaling-up problem." Essentially this problem occurs because biological organisms in large quantities behave differently than chemicals. The small biotech firms had to turn to the pharmaceutical firms for this technological expertise and in addition to pay for a clinical trial, which can cost as much as $1 billion, because they had the deep pockets. Finally, the large drug companies also had expertise in how best to commercialize a product once the FDA accepted it. The interesting point is that pharmaceutical companies could not start their own biotech divisions because their thinking about research problems was too different, having an emphasis on chemical solutions such as drugs rather than on biological solutions. Hence, to have access to these new developments, the drug companies either formed joint ventures with particular biotech firms or purchased them. Via these practices, the large drug companies could monitor developments in these new areas of medical research. Hence networks of various kinds formed between the small biotech firms and the larger pharmaceutical companies as well as between the former and universities. Previously, we suggested that evolution was producing new opportunities for cooperation. The patterns described in Figure 1.3 are examples, but they are not the only ones. Although this evolutionary pattern in biotech appears to be straightforward, it did not occur everywhere.

Figure 1.3 can also be used to define and illustrate the valley of death. Its definition is the failure to transfer basic and applied scientific research into commercialized products. The dotted lines between the national laboratories, most particularly the National Institutes of Health (NIH), reflect the lack of translation of their basic and applied research into products. The dotted arrows reflect intramural research, for example, the large program on cancer research, whereas the arrow from public research laboratories (e.g., NIH) to the universities represents the extramural program. (As indicated in Figure I.2, we have a deficit in the trade balance for health care products despite the very large expenditures on medical research.) The steadily smaller arrows in Figure 1.3 between the public research laboratories and the universities, between the universities and the biotech companies, and between the biotech companies and the pharmaceutical firms also indicate that this alternative way of moving ideas from public research laboratories such as NIH into commercialization is not strong enough.

One of the great advantages of this evolutionary pattern is that it allows both managers and policy makers to detect where failed evolution occurs, that is, where one of the predictions of the theory does not unfold. The model of evolution adds to a number of academic topics which are discussed in the concluding sections of each chapter. Failed evolution may occur in any of the following dimensions:[19]

1. Differentiation of research arenas with separate organizations
2. Creation of new research organizations, frequently small ones, within some of them
3. Formation of networks within and between the differentiated research arenas
4. Strengthening of the connections in these networks

In France and Germany, various obstacles slowed this evolutionary process in the biotechnology and pharmaceutical sector, especially the first two areas in the given list. To correct for this "evolutionary failure," the German government stepped in with a series of policy remedies to facilitate the creation of a small biotech sector in Munich with some success.[20] But they have not eliminated all the barriers because they have been less successful in stimulating radical innovations in this sector. *One of the themes of this book is the important role that policy makers can play by intervening to eliminate obstacles in the evolutionary patterns previously predicted.* For this reason, Chapter 6 provides measures of the radicalness of technological advance for each arena so that policy makers can easily detect where failure appears to be.

A major obstacle for managers and policy makers is the lack of integration between arenas once they become differentiated. These produce gaps in the transmission of technical advances from one arena to another. The central tenet of this evolutionary innovation process is that for radical product or process innovation to occur rapidly, the differentiated arenas have to be connected tightly so that knowledge advances can be easily monitored. The transmission of advances is more difficult than it might appear at first glance, and the more radical the advance, the more difficult it becomes. By definition, a radical advance is "outside the box" and therefore not necessarily easy to understand within the prevailing ways of thinking within either the scientific or engineering specialty or management practice. Thus both the communication of the advance and its comprehension are not straightforward. Relying upon the market and the usual venues of transmission (conferences, journals, newspapers, the Internet) and assuming that the information is not proprietary does not provide the necessary background information about the assumptions and methods that undergird the technical advance, assumptions and methods that are frequently difficult for the researcher alone to verbalize.[21] Only intense and frequent interaction can lead to the kind of understanding that allows the technical advance to have an import in another arena. Hence, the concern about the tightness of the connection between arenas: being able to understand the technical advance well enough to ask the right questions about the underlying assumptions and methods. *One of the reasons why radical product and process innovation tends to take so long is because of the lack of frequent and intense interaction between the different research arenas that have become differentiated into separate research organizations and firms and even between the different arenas within the same firm, known as "stovepipes."* (See Chapters 4 and 5.)

The recently published book *Open Innovation* provides some evidence for the evolutionary patterns discussed in this chapter.[22] Essentially, the book argues that the closed model of innovation, in which all six arenas are contained within the same organization's industrial laboratory, is no longer viable. It reports an extensive case study of Xerox PARC (subsequently changed to the Palo Alto Research Center), the amazing research laboratory. However, the book does not indicate how to coordinate the information about technical advances in each of the six arenas or how this might vary from one industrial sector to the next. The perils of open innovation are the failure to connect quickly and with enough strength so that new scientific breakthroughs are quickly exploited. Indeed, there is little about scientific breakthroughs in the book. With tight connections in the idea innovation network, radical innovations can occur much more rapidly, despite the commonsense notion mentioned in the Introduction that they should take longer because of the differentiation of the arenas. Hence, in various chapters, considerable emphasis is placed on the importance of frequent and intense interaction or strong connectedness between differentiated arenas in the networks that connect them and, most especially, the connections between basic and applied research on the one hand and manufacturing and quality research on the other. To create this tightness, however, implies cooperation between the public and private sectors, a level of cooperation that is sorely missing at present.

Because understanding this evolutionary pattern is so critical to restoring the innovative edge and, as we have seen in the Introduction, is the basis for a new policy paradigm built around the concepts of evolution and failed evolution, it is worth establishing the evidence for this pattern. Besides the research documenting Figure 1.3 for the American pharmaceutical and biotech sector and the interventions of the German policy makers to duplicate the American pattern, there is a considerable amount of evidence in different and disparate literatures. Perhaps the most interesting piece of evidence for this impact is the gradual withdrawal of major American companies from basic research relative to their investments in applied research and product development, as we observed in the Introduction. A very different set of studies, which focuses on the scientific and applied research arenas rather than on the manufacturing and quality research arenas, argues for the spread of networks around universities. But this argument only applies to those technological sectors, such as biotech and pharmaceutical companies, where expenditures for applied research are relatively small, and not to a sector such as aircraft or automobile construction.[23]

Three additional pieces of evidence are provided to support the evolution in how innovation is produced. First, the reasons why evolutionary differentiation proceeds should be indicated because it can make the argument more plausible and help answer an important question: Why would large companies with successful histories of industrial research, in their own laboratories, reduce their expenditures in basic research? We discuss the reasons for the apparent paradox. Second, an intensive study of two high-tech sectors in Europe provides further evidence and at the same time indicates how different this evolutionary pattern is across high-tech sectors. Third, in Chapters 4, 5, and 6, additional evidence is provided relative to different sectors, economic and noneconomic.

This chapter started with the question: Can the future be predicted? To have some insights about when some of the research arenas depicted in Table I.1 move out of the industrial laboratory and become differentiated in separate organizations necessitates understanding the processes of how new organizations emerge in differentiated research arenas as well as the parallel process of the creation of new occupational specialties.

The growth in knowledge, as reflected by the rise of RDT expenditures, produces both occupational specialization and differentiation of research organizations into separate functional arenas. The driving force behind the need for specialization is the need for greater focus. At the individual level, there is a limit to how much new information can be absorbed, and at the organizational level, the limits are raised by the amount of communication between the different research teams. As the diversity of different topics grows, it requires more and more communication, which necessitates effort, as we will see in Chapter 3. As the amount of expenditures within a particular functional arena grows, it becomes useful to differentiate that arena because of the advantages of greater focus produced by specialization. In other words, this differentiation parallels the creation of new scientific disciplines from time to time. The functional specialization allows for greater focus. Organizations have very definite limits in coordinating a diverse array of specialties, especially if one desires intense and frequent interaction, which is necessary for the creation of true understanding, that is, the hidden assumptions and methods involved in making a radical advance.[24] At various points, I have mentioned the problem of "stovepipes," that is, the breakdown in communication between disparate research arenas within the same organization. Furthermore, once specialization occurs, it becomes more possible to observe the relevance of other disciplines or applied areas of research. Another advantage of specialization is that, paradoxically, this greater focus allows for quicker solutions to problems. This advantage of specialization is one of the main reasons why the large high-tech companies have been moving away from basic research, examples of which are provided in Chapter 5. They cannot quickly exploit the fruits of their own basic research. Furthermore, to do so frequently requires a new kind of business plan, as the analysis of Xerox PARC makes so apparent.[25] Why developing a new business plan is difficult is explained later in this chapter.

But given specialization and the need for expertise in other arenas, even in other sectors, some form of integration within research teams and across organizational boundaries becomes necessary. Integration tends to be free forming, but when it does not occur or if the integration develops slowly, then an evolutionary failure occurs that manifests itself in a decline in the rate of radical product and process innovation in that sector. Furthermore, networks or interorganizational relationships vary a great deal in both size and global reach; sometimes they are simply knowledge communities or research consortia within countries, sometimes organizations form international interorganizational relationships, and at other times they establish formal global alliances for radical product development.[26]

One important kind of evolutionary failure or obstacle to restoring the innovative edge is *weak connections* between the arenas and, in particular, between those in the public sector and those in the private sector. Thus the United States does not exploit as much of its basic research in the national re-

search laboratories as it could. When scientific breakthroughs or technological leaps take place, there is much that is not understood, and only frequent
and intense interaction will provide a strong enough connection so that the
new knowledge can be transmitted successfully to the other arenas. Implied
in the idea of the necessity for a strong connection is a preference for the
interorganizational networks as a coordination mode that overcomes communication gaps created by the advance.

Also implicit and perhaps more interesting is the central insight previously mentioned of ideas moving back and forth in a network and its implications for the design and management of the idea innovation network.[27] As
organizational and occupational differentiation unfolds, the only way in
which the network of ideas is maintained is if all arenas are connected to each
other in some way. If one moves away from the simple notion of connecting
basic and applied research or manufacturing with qualities research to combining basic with manufacturing research or applied with qualities research,
then even more of a communication gap is created, and the barriers to understanding the hidden assumptions and methods become greater as well. This
communication gap represents a significant kind of obstacle, a topic that we
return to in Chapter 3. Another reason why large research organizations and
large firms have difficulty in learning about scientific breakthroughs or technical leaps is the negative influence of their organizational culture or typical
ways of thinking about problems. Organizations, especially successful ones,
have developed a sense of what is important and routines or procedures,
including ways of thinking, about how to conduct research. For example, in
pharmaceutical companies organizational cultures have been built around
chemistry, a discipline which is distinct from that of biology; hence, the companies encounter difficulty in managing biotech divisions and require differentiation. To absorb radically new products and manufacturing processes, the
cultures of organizations have to change. Scientists with new ideas about gene
therapy or some other treatment built on the discipline of biology find it much
easier to start their own company rather than work in a pharmaceutical firm,
especially given the availability of venture capital. Hence, absorbing new ideas
from the environment is much more difficult than the concept of open innovation would lead one to believe.[28] For example, perhaps the most important
reason why Xerox did not profit more from its radical innovations, such as over-
lapping windows, menus, screen icons, and the mouse, which were developed
in its research organization PARC, was that the company had an engineering

culture that was not open to alternative technologies.[29] As a consequence, some of the most creative individuals went to Apple. As I have already indicated, this is one of the larger messages about Xerox PARC; it only kept those technologies that fit its business model. Management could have sought expertise from specialized firms to develop a business plan for each new technology and then have created a separate organization to pursue the new technology. Even better would have been the creation of complex research teams that had the needed expertise for commercializing the new ideas in a different technological sector. For example, P&G did this, working closely with its two major distributors, Target and Wal-Mart, and learning how to better distribute its products.[30]

What is especially unusual about this evolutionary pattern, as already observed in the Introduction, is the implicit argument, at least in the high-tech sectors, that one needs to have specialized functional arenas of research not only in basic and applied research, which are typical, but also in product development, manufacturing research, quality research including the reduction of externalities, and commercialization research. It is the amplitude of the differentiation that is predicted or the variety of ways in which scientific and technological research can be added to the research teams and networks that are developing innovative products and processes that represents a major obstacle for both managers and policy makers.

What are the advantages of this model for policy makers in the debate about "picking winners"? The model of failed evolution as defined by the lack of movement in a specific sector arena along one or more of the four dimensions previously outlined in this section provides policy makers with a much more refined tool for searching for obstacles that need to be overcome. Rather than arguing that one should support this or that research topic (for instance, nanotechnology), or this or that industry (e.g., automobiles), it asks where the failure is. Is some arena totally ignored (e.g., manufacturing research or quality research)? Is the failure a lack of differentiation of specialized research arenas because of the absence of new small high-tech companies, and if so, does policy need to stimulate small high-tech companies as Germany did? Are there communication gaps between arenas, in particular between the scientific research located in the national laboratories and the manufacturing arenas of the business firms, and thus a "valley of death" policy problem? Is it because of weak connections between one or more arenas, and therefore the communication is neither effective nor quick enough? This ideal model, which must be adapted to each high-tech sector, as indicated later in this chapter, makes a series of predictions. If these predictions do not occur, then the failures in evolution have to be corrected by policy makers or transformational CEOs. Although there are many exceptions, the evolutionary failures in the United States tend to be caused by the lack of development in networks to connect the differentiated arenas of research, whereas the failures in Europe tend to be the lack of emergence of small high-tech companies willing to take high risks.[31]

The predictions contained in Figure 1.3 have more support than just the example of the American pharmaceutical and biotech industry. They apply to other high-technological sectors and other parts of the world. With funding from the European Union, four research teams, one each for Austria, Finland, Germany, and the Netherlands, tested the theory.[32] The test involved contrasting and comparing the two high-tech sectors of telecommunications and pharmaceuticals.[33] Each research team knew the language, the culture, and the history of the country, and therefore could more quickly and accurately study the evolutionary trends that occurred in the idea innovation networks within each of these two sectors as a consequence of either a major set of scientific advances (biotech) or a number of technological breakthroughs (telecommunications), particularly over the period 1985–2000. What is especially interesting about the team study in the pharmaceutical case is that the radical advances were paradigmatic breakthroughs in basic science (molecular biology, genetics) and applied research and especially in the development of a number of new tools grouped around the idea of biotechnology, making applied research and product development important arenas. In contrast, the radical advances in the telecommunications case represented the development of a number of technological breakthroughs, including digital rather than analog transmission, the movement to network architecture, and the use of optical fiber cables that could handle large volumes of data transmission, making both quality and marketing research the more critical of the six arenas.[34] Some of the superior performance characteristics of digital in comparison to analog transmission are easier noise filtration, higher speed, better voice quality, and fewer errors. Likewise, the superior qualities of optic fibers, besides the higher volume of transmission, include that they are more difficult to tap illegally and have lower operating and maintenance costs, gains in quality in multiple ways.

Although the national teams focused primarily on the last fifteen years, they established the historical context for each of these two sectors. In the beginning, the typical telecommunications company was actually a public sector monopoly that handled both the telephone and postal service in the four European countries. (In the United States, it was a private company for the telephone, AT&T, and a public service for the post office.) The equipment suppliers were also monopolies but not the same public company as the service provider (unlike in the United States, where AT&T was also the equipment supplier, Western Electric). There were no functionally differentiated arenas because all of the research was in-house both in Europe and in the United States (the famous Bell Laboratories).

The new technologies allowed for new products such as mobile phones, and new forms of telecommunication networks that created opportunities for new service providers. Part of the reason why there were new companies was because of a change in the institutional rules of the game; both Europe and the United States moved away from monopolies and allowed market liberalization.[35] Although a number of new firms for services and equipment emerged after a period of time in Europe, a new concentration was established because of the large capital costs of the networks for services. However, the new concentration became international, as companies in Europe bought out their competitors in other countries. As a consequence, now two of the countries host major international players, Finland with Nokia and German with Siemens, while the other countries only have companies in niche markets, the Netherlands in cables and Austria in speech processing for Eastern Europe.

Despite these trends, the new concentration did not imply there was a lack of space for small companies because, increasingly, the major players outsourced various activities. In particular, the telecommunications industry withdrew from basic and applied research. These arenas moved into the universities and specialized research institutes, depending upon the particular country and its historical patterns. As the idea innovation network evolutionary model predicts, functional differentiation of research did occur. There is also now much more cooperation with researchers from different companies who join forces in particular projects and even across national boundaries. And while many of these networks are informal, they are also constructed with formal contracts to protect property rights.

The increases in the complexity of the idea innovation network were particularly striking in the telecommunications sector.[36] The improvement from the combination of digital, network structure, and optic fibers led to a proliferation of end products, linkages between different systems, and the blurring of the sector boundaries, especially between computer/science, telecommunications, and multimedia, another illustration of the evolution toward greater complexity. Industrial products have become sectors in themselves, for example, mobile phones. This blurring is less apparent in biotech but is happening there as well.

Another prediction of the model of evolution is the development of networks connecting smaller organizations with larger ones. Subnetworks developed in the following areas: specialized research, marketing, physical infrastructure of the networks, manufacturing of end-line equipment, and managing of call centers. A particularly interesting development is the telecommunications manufacturers organizing mobile operating and business equipment, and customers in international user groups where networks of product managers, marketers, users, and researchers discuss problems. In other words, commercialization research has become an extremely important part of the strategy for product innovation.

Much the same story can be told about the biotech and pharmaceutical sector. The pattern is less sharp, however, because this story involves three industries, namely, agriculture, drugs, and environmental protection. Concentration remains because of the very large development costs for clinical trials and because of marketing for new drugs. But again subnetworks of small organizations have emerged in research and product development. Again we observe functional differentiation of arenas. But there is a difference between the two sectors. In telecommunications, they are midstream and downstream research arenas that evolve, whereas in pharmaceuticals they are upstream arenas. Again, these examples illustrate the importance of studying each technological sector. One observes different patterns in the processes of differentiation and the potential for evolutionary failures either in the emergence of new high-tech companies or networks to bridge the gaps in the idea innovation network that develop.

One of the more interesting findings of this study is that the increasing importance of science for product and process innovation has led to the research arena of basic science becoming more differentiated. This finding substantiates perhaps one of the most important insights of the evolutionary model of the idea innovation network. In the past, the idea innovation chains were mainly perceived to be a question of product development. Now, more and more, product development requires a good scientific base. A common finding in the comparison between the two sectors is that radical technical advances can sometimes produce discontinuous change, which makes certain competencies obsolete or at minimum requires new competencies, which is why they are sometimes called disruptive technologies.[37]

Another important finding is that the main innovation networks have been differentiated into subnetworks within specific arenas that have become interdependent as well, creating a complex web of networks within each technological sector. A parallel process has happened with the supply chains of the major equipment suppliers who have surrounded themselves with many suppliers and have conducted joint research with them. In telecommunications, the major players are outsourcing manufacturing, starting with accessories and parts, then whole products, and finally even the assembly work. The extreme case is Nokia, which now specializes in research, software production, product design, and brand management. Actual production is outsourced, and 30 percent of the Nokia workforce worldwide performs research in various arenas. A remarkable variety of hybrid organizations have emerged to form the needed integration and to access the hidden assumptions and methods behind radical advances, including joint ventures, user groups, product teams, patent pools, collective trademarks, technology clusters, partnerships, alliances, and even virtual firms. In other words, network coordination can be implemented in a variety of ways.

Finally, the European study reports that increasingly idea innovation networks are international in scope even with specialization by country.[38] Most of this specialization is occurring either at the beginning of the idea innovation network or at the end. They observe an important implication of this trend: Precisely because these sectors are becoming internationalized, the limits of the country-specific institutions are becoming less restrictive.[39] In particular, European, especially German, pharmaceutical companies have been investing in or buying American biotech firms.[40] The large companies can hire anywhere in the world. Most of the new start-ups in biotech in Germany were founded by scientists who had been American trained.[41] Likewise, one can escape various regulation restrictions on research, such as stem cell research, by outsourcing the research to countries that do not have regulatory restrictions, as American firms are doing.

Given this strong evidence for this evolutionary pattern, managers of research organizations and of firms concerned about stimulating radical product and process innovation need to increase the scientific quotient in their products. But rather than simply regarding this goal as an issue of basic research, the idea innovation network model broadens the concept of scientific and technological research to include six kinds of research (BAPMQC) that have to be considered in the development of radical innovations. Rather than restrict research to the usual trilogy of basic, applied, and product development, the idea innovation network model stresses the equal importance of manufacturing research, quality research, and commercialization research, that is, the development of new kinds of business plans. In the next two sections, we also consider the importance of manufacturing research, an arena that is frequently neglected in discussions about restoring the innovation edge.[42]

Another critical implication is that, in order to increase the scientific and technological quotient in the development of radically new products and processes, managers of research organizations and of business organizations have to develop networks to close the gaps of communication. How gaps are closed varies across technological sectors. Clearly, the patterns of networks were different in the two technological sectors that were compared in the reported European test. Unlike the simple prescription of open innovation, managers should consider how their specific technological sector is evolving and therefore where networks have to be constituted so that the researchers keep abreast of cutting-edge research findings that impact on their own specific arena.

The third implication that follows from the previous one is that cooperation is increasingly necessary within the idea innovation network, if not also across several idea innovation networks, as products are created with many different kinds of components. The same is true for the supply chains attached to the major components in complex products. One business firm cannot handle these many different kinds of specialized research by itself; it needs partners to do this. Furthermore, as I shall discuss in Chapter 5, cooperation between competitors within the same idea innovation network is also increasingly becoming both more common and more necessary as various national technological sectors compete against their counterparts in other countries. The semiconductor industry in the United States is an example of how cooperation can be accomplished.

But the changing rules of competition, from fierce competition to cooperation, at least with some competitors and with other organizations within the same idea innovation network, are just a consequence of the need to build networks. Another force that is encouraging cooperation is the steady proliferation of market niches as various companies and research organizations attempt to meet the variety of demands for different kinds of products and services. In other words, this evolution in the nature of competition flows from the evolution in the nature of tastes toward more and more customization.

THE EVOLUTION IN CONCERNS

Education, as I have previously described, not only plays a role in the evolution of tastes toward more technological sophistication, higher quality, good design, and individualized tastes but also helps explain the increasing concerns of the consumer about the impact of both the manufacturing of the product and the use of the product on the environment, the hidden costs that we mentioned in the Introduction. These concerns have led to a number of social movements and organizations that attempt to raise public consciousness about various issues. The Sierra Club, Greenpeace, Clean Water Action, and the World Wildlife Fund are perhaps some of the best known. More recently, a number of new social movements have emerged to fight for corporate responsibility. But these are not the only issues. An overemphasis on the state of the economy and economic growth obscures perhaps the most interesting fact about the dawn of the twenty-first century, which is the emergence of new global problems that were present previously but became better recognized because of various dramatic incidents: 9/11, the struggle between the Sunni and the Shi'ia in Iraq, the disappearance of arctic ice and the placing of the polar bear on the endangered species list, the explosion in the cost of primary metals and other nonrenewable resources, and the appearance of avian flu, to say nothing about AIDS.

But dramatic incidents, while riveting our attention, are not the only ways in which our awareness of these global problems has increased. Let us not forget the special role of scientific research on the hidden costs of manufacturing. In particular, as research has deepened we have also come to understand better the ill effects of carbon dioxide on our environment, the role of asbestos and lead on our health, the impacts of cell phones on our brains, and so on. Each year the public learns once again that there are hidden costs in the products and services that we use. And even though drugs and surgical techniques are designed to treat illnesses and to save lives, there are frequently long-term costs that only become apparent after many years of research on specific health techniques. The recent controversy about the frequency and the appropriate age for the beginning of screening for breast cancer with mammograms illustrates the complex issues involved.

Probably everyone has their own list of what these new global strategies are, but I think, at minimum, most would accept the following:

1. Increase homeland security via the creation of national security sensors and training of personnel for protection against terrorist attacks.
2. Develop ways of handling tribal warfare via the creation of new kinds of military equipment and the training of officers and personnel to handle tribal and civil wars as well peacekeeping.
3. Reduce global warming (carbon footprints) and develop alternative energy sources besides fossil fuels.
4. Reduce the use of nonrenewable resources with the development of new materials via nanotechnologies.
5. Prevent and treat pandemics quickly.

Each of these new global strategies requires not only the development of new technologies and services but the retraining of various kinds of personnel to help operate these new technologies and provide their attached services. Checking for weapons as part of airport security is a good metaphor for the kinds of technologies that have to be developed and the specialized training that should accompany these new technologies. And as this example makes clear, scientific research has to be an important part of the solution. An interesting example of what basic and applied research can accomplish is the development of check-in monitors for trains and planes that can detect small amounts of explosive material, research being conducted at one of DOE's national research laboratories.

In some sense, each of these objectives is now part of existing programs of research. But precisely because they are scattered across several sectors or represent a niche within one of the eleven high-tech sectors, and more critically involve different government agencies, there is perhaps not enough recognition of them as distinct and separate noneconomic technological sectors. One of my recommendations would be for the federal government to create new technological sectors focused on the resolution of these five global problems. At minimum, the federal and state governments should start collecting data on each noneconomic sector that corresponds to some national goals such as those suggested in the previous list—the number of research organizations, firms, and the like—in the same way they presently collect data on various economic sectors. In fact, each of these sectors has products and services that can be sold for profit. And consistent with the idea of our society being a service society, in these areas services are increasingly an extremely important part of the product, including the training of individuals in how to use these products. Finally, by viewing these areas as relevant sectors in addition to the list of the eleven high-tech sectors allows for the development of indicators of success in dealing with these problems, itself an important way of focusing attention.

Once one stipulates research related to solving global problems as new technological sectors, three major advantages occur. First, we observe that there are many opportunities for creating radical products: new national security sensors, military equipment for tribal wars, alternative energies, and new materials to substitute for nonrenewable resources, as well as treatments for worldwide epidemics. In four of these areas, it implies a considerable range of products, technologies, and services. Just think of all the different kinds of potential terrorist targets that need to be protected, including dams, water purification plants, important buildings, subways, and trains, as well as airports. Second, viewing the resolution of global problems as technological sectors focuses attention on the different kinds of BAPMQC investments, as outlined in Table I.1, needed to produce these new products, technologies, and services. The same logic used there can be applied to these new technological sectors defined by national missions because they involve both the sale of products/services and investments in different kinds of research. An argument made later in this chapter is that to reduce carbon footprints and have a significant impact on global warming necessitates changing the way all products and services are produced, a tall order but one that generates many radical process innovations that can reduce the negative trade balance. Third, collecting data on these new technological sectors and treating them as distinct not only leads to an understanding of new ways of helping our balance of payment problems but should lead to establishing a clear output relative to a political goal such as reduction in terrorist attacks, success in tribal warfare, or decrease in carbon footprints, creating a new form of thinking about government responsibilities. Everyone wins— increased employment, profits, and tax revenues—when the five goals listed previously are partially achieved. Later in this chapter, I will discuss only the issue of reducing carbon footprints, because it allows for such a massive attack on our trade balance problem and across all economic sectors. Also, it is obviously tied to the issue of dependency upon oil and its implications for entanglements in foreign wars, as well as to the domestic concern about offshore drilling and oil spills, the hidden cost of which has become dramatically visible.

The problems of global warming present many opportunities for the creation of both radically new products and manufacturing processes, both radical product and process innovation.[43] The former has been the most emphasized in the discussions of alternative energies as new products such as windmills, solar panels, nuclear energy, and ethanol fuels. Fortunately, a considerable amount of RDT funds, about $15 billion including the stimulus funds, has been invested in DOE to tackle these and other issues relative to energy. Unfortunately, the energy sector is again where the United States lags technologically despite the programs within DOE, with the best fuel cells, wind technology, and energy storage having been developed outside the United States.[44] Denmark now controls one-third of the global market for wind turbines, and Germany is making a concerted effort in this area as well. France has perfected a number of the Westinghouse patents relative to nuclear energy and is now in a position to sell these products worldwide. In contrast, in the area of photovoltaics U.S. research is making great strides in power-generating capacity (see example in Chapter 6), which is increasing exponentially. However, developing more efficient thermal couplings to transfer the heat so that it is converted into energy now needs to be a priority.

But the problems are quite complex and illustrate in various ways the importance of having not only basic and applied research but research in the other arenas as well, in particular on how to create distribution systems for the new forms of energy. Consider the case of the hydrogen-powered car, for which the following radical innovations have to be developed: inexpensive and long-lasting fuel cells, an adequate fuel shortage capacity at a reasonable price, a carbon-free way of generating the hydrogen, and a national refueling system.[45] Nor is the hydrogen fuel cell the panacea for efficient automobile transportation. What is now increasingly recognized is that the principle of customization applies to cars as well as other modes of transportation. Different vehicles will need different kinds of energy sources. Energy efficiency shifts depending upon the size of the car, SUV, or truck.[46]

Again, the need for research on distribution of alternative energies must not be forgotten. The problem is how to construct tomorrow's electrical grid so that it can incorporate wind and solar power seamlessly and reliably.[47] But in each of these alternative energies, there are great opportunities if the right kind of scientific and technological research is brought to bear.

The discussion of new processes that reduce carbon footprints has received less attention despite a program in the DOE concerned with energy efficiency in the manufacturing of products. Within this perspective, the proposal contained in N. R. Augustine's *Rising Above the Gathering Storm* (2005) to create a special research funding organization, Advanced Research Projects Agency–Energy (ARPA-E), has been included in the stimulus package and been given $400 million as of 2009. However, how effectively this money will be spent depends upon how this organization is managed. If a transformational leadership team along the lines discussed in Chapter 4 is created, then more radical innovations are likely.

The focus on creating manufacturing systems that utilize little energy and products that do not use much energy during operation represents an opportunity for the United States to take a major lead in many of the traditional industries where we have lost considerable market share. The efforts in reducing the consumption of energy in cars are visible in the increased average gas mileage of the U.S. fleet of cars. Less visible is the need for developing new technologies for the *re*construction of existing homes, plants, and buildings so that they utilize much less energy. Supporting the refitting of the present capital stock of the country can have an enormous impact on the amount of carbon produced in this country and, I might add, on the economy. Furthermore, developing prefab solutions for green building construction that can be sold to developed countries, if not also to developing countries, could reduce global energy consumption.

In considering research projects for ARPA-E, some attention should be paid to the consequences of global warming for the United States and how to adapt to these consequences. Even if carbon footprints are stabilized within the next two decades, which seems doubtful, the impact of the current levels will cause continued global warming for at least several decades beyond the period of stabilization. Preparing for these impacts leaves open a number of new opportunities for research in agriculture, fishing, and public works construction to handle floods in the various rivers of the United States and to protect the coasts from more severe hurricanes.

Some of the opportunities for new areas of agricultural and oceanographic research emerge from some of the contradictory consequences of global warming: too much water in some places and not enough in others; the steady heating of the oceans and its impact on fish stocks; and more critically, the impact of warming on the frequency and intensity of violent storms of different kinds (tornadoes, hurricanes, cyclones). Furthermore, the effects of global warming are not directly linear but have complex feedbacks that have to be specified, which is one reason why some continue to deny that there is global warming or that mankind causes it. For example, the heating of the oceans leads to more intense hurricanes, but it also means more wind shear, which dampens the intensity of the hurricanes. The more critical issue is not the basic research on the causes of hurricanes but how to prepare for intense ones, such as Hurricane Katrina, when they do occur. Whole new kinds of flood control have to be developed, and how buildings can be better protected against violent winds that are associated with tornadoes and hurricanes needs to be reevaluated.

A good example of the contradictory consequences of global warming and attempts to deal with them can be seen in agriculture. In some agricultural areas, there is too little water, requiring the development of food products that can survive extended periods of drought. One observes some movement in this direction as agricultural research attempts to develop crops that can grow without much water and also crops with increased yields. Another example is the development of ethanol fuels to reduce dependency upon oil, which has resulted in a shortage of grains for global food consumption. Clearly, research on intensive agriculture methods that can adapt to long periods of dryness is required.[48] But on the other side, we need to develop new kinds of crops that can withstand too much rain, especially at particular moments—a problem that manifested itself in 2008.

The warming of the oceans is leading to the destruction of coral reefs, which are the breeding grounds for one-fourth of our fish stocks, and of oyster beds—as in the Chesapeake Bay—as well as to the creation of harmful algae blooms there and elsewhere. Dealing with these consequences requires a lot of creative research, some of which is occurring. But the larger issue of how to restore fish stocks in the oceans, with other countries competing for this resource, necessitates a number of innovations in the management of fish stocks done cooperatively. Worth mentioning is the recent breakthrough in the farming of tuna, a prized fish for many countries.[49]

Research on how to protect people who live in the areas where tornadoes are common has clearly become critical, given their recent increases in frequency. And unlike hurricanes, as yet there do not appear to be countervailing factors that will reduce their frequency and intensity. Among other needs, this requires rethinking of building codes and developing more effective warning systems.

If drought is a problem in some parts of the United States, floods are also becoming more frequent in other areas. The whole system of levees and protection against floods has to be rethought. Again, the issue is not just building something bigger and stronger but creating systems that are much more flexible in handling these extremes.

One could go on, but it should be reasonably clear that the opportunities for innovative solutions to the problems of global warming are almost endless. Being able to create export industries in alternative energies and in solutions to the problems of global warming would strengthen the hand of the United States as it attempts to convince the developing countries of the necessity of reducing their carbon footprints at the same time we are reducing ours. We have to show them how this can be done and use the export of capable products and process as a bargaining chip in the negotiations.

THE CONTINUED IMPORTANCE OF PRODUCTIVITY

Parallel with the globalization of work is the globalization of demand for products. Rather than export jobs, let us export goods and services designed to meet the cultural needs of other countries. The distinctive characteristic of this globalization is the increasing diversity of tastes relative to a number of technological issues, quality, and design, but at the same time, the differences are not always as great as one would imagine when internal variations by class and region are considered. The disparities in tastes between what the Chinese and the Canadians would like to see in the same product are not obvious and need to be researched. The question then becomes the following: What kind of car does the middle-class Chinese person living in Shanghai want in comparison to the middle-class Canadian living in Toronto, to say nothing about the difference with her counterpart in Montreal. As a consequence, the desired quality in the product and service, the relative importance of technological advances, and the specific kind of design vary less across than within countries. To maintain productivity in world markets with diverse tastes, manufacturers have to master the production of many different kinds of product attributes on the same assembly line in small batches that can be quickly changed. As reported in the Introduction, the United States succeeded in having a rapid economic development because of its mastery of the first divide in manufacturing, the assembly line, which is sometimes called Fordism, that moved away from small-batch production to mass production. But the United States, with some exceptions, has failed in the adoption of flexible manufacturing, which is called post-Fordism and is the second divide, as we saw in the Introduction.

Despite this failure, the importance of flexible manufacturing has increased even more because of the evolutionary pattern of customization and also because of globalization. In addition to handling a variety of tastes, American manufacturers have to worry about steadily improving the quality of the products that they produce and being able to quickly change their production lines to handle the different niches in the market. The Swiss manufacturer of instant coffee, Nescafe, has refined its manufacturing processes to produce distinctive flavors of coffee even for different regions of the United States. The Japanese, in particular, have been successful in constantly improving the quality of their products, such as cars, although the recent recalls of Toyota indicate that failure is always possible, particularly when expansion in production is too fast. Being first with the latest technological edge and with good design are also increasingly important in a wide array of consumer products from Mac laptops to Italian espresso machines to German cars.

But implementing flexible manufacturing may not be enough. I think we may need a *third* industrial divide. Not only should manufacturing machines and techniques be reprogrammable consistent with the thesis of post-Fordism, but they should also be designed and built so that they can easily absorb the advances in the new materials science or nanotechnology. The third industrial divide would involve another level of flexibility, the ability to change the *composition* of the materials used *to make* the components of the parts employed in the construction of a product, and not just the shape of the components, which is the second industrial divide. Besides continually developing the manufacturing and service provision capacity to customize products and services, we also need to continually develop the capacity to customize their supply chains. In one sense, this means adapting the idea of process technologies, such as are used in chemical plants and in oil refining installations, to the making of many other kinds of products, and the provision of services. Some of the exciting potential is hinted at by the use of chemical molecules in developing new kinds of computer processors.

The manufacturing of the car represents what might be an interesting example of how one might advance to the third industrial divide. More and more the amount of steel is being reduced and replaced with other kinds of materials, typically plastics designed to have certain properties. Going back to the time of the Model T, when Ford had a steel plant, the River Rouge, constructed alongside the assembly plant for this car, the supply chain was completely integrated into the construction of automobile parts and their assembly. This kind of integration makes a great deal of economic sense when standardized parts are being manufactured over long time periods; the Model T lasted in production for approximately eighteen years. But of course when car manufacturers, especially GM, moved in the direction of different kinds of cars for different kinds of people, then the supply chain became more separated from the assembly of the car. As we move toward customization of cars, one can imagine the desirability of being able to customize the properties of the parts that constitute the car in the assembly plant. In particular, the substitution of plastics, and perhaps even more interestingly, the combination of plastics with glass designed to have superior safety features in crashes, raises the possibility of again thinking about the integration of the supply chain into the production of parts to be assembled in the car. The example of substituting an "alloy" of plastics and glass in the construction of the car also makes more possible process technologies for the composition of components for the car, and thus the third industrial divide becomes possible because these components can be conceived as fluids.

The integration of the supply chain with the car assembly only makes *scientific* sense if the basic and applied science and the manufacturing research develop economical ways of accomplishing this integration with process technologies. And it only makes *economic* sense if (1) assembly lines handle a broader and broader range of cars and trucks for disparate customers that necessitate different kinds of components with distinctive material properties, and (2) the energy costs of transportation stay high, as they likely will, because of the growing demands from the developing countries for fossil fuels.

An example of one area where the third industrial divide exists is the company Modumetal, a small high-tech firm in Seattle.[50] Scraps of nickel, iron, and other metals are dissolved in acid and their atomic structures are then revised via a series of electrical shocks from a conducting rod in the acid bath. The new alloy grows onto ultra-thin layers one nanometer (one billionth of a meter) thick within a mold. Electrical shocks are also used to create layers as they grow "from the ground up." By combining layers, this "Frankenstein" alloy can become resistant to shattering and penetration, and at the same time have lightweight, metallic properties that do not occur together naturally. The company describes this metal as plywood with similar advantages. The process the company uses can make one piece of metal harder in one area and softer in another, and perhaps more critically, stronger than steel.

These new alloys can be used to make military body armor, currently 32 pounds, about half as heavy, reducing the fatigue of the soldier, increasing his mobility while providing better protection. Likewise, the reduction in weight of the protective armor of military vehicles, such as Humvees (High-Mobility Multipurpose Wheeled Vehicles), has enormous impact on gasoline consumption, the speed of the vehicle, its maneuverability, and a number of other desirable properties.

The third industrial divide may not be either practical or even possible for large-scale production, but it is clearly able to handle small batches of products. The emphasis on being able to change the composition of the components in products, if successfully accomplished on a larger scale and in different sectors, provides ways in which manufacturing can be returned to the United States.

One of the secondary themes in this book is how to restore our traditional industries. In particular, finding the right kinds of manufacturing equipment that can reduce global warming, the use of nonrenewable materials, and energy consumption in the manufacturing of products provide opportunities especially if these are customized to the specific needs of particular nations. If we can build machines that reduce hidden costs to individuals and the environment in the manufacturing of paper, toys, textiles, and other low-tech products, then we can stimulate and revitalize our tool and die industry. One of the ways in which Europe in general and Germany in particular maintain exports to the United States and elsewhere is the quality of the machines that it sells to various countries. For example, the Italians sell the automated machinery for producing wine, pressing olive oil, and processing of tomato products. Sweden has excelled in creating highly "green" production of wood pulp for making paper products. The Germans export highly sophisticated machinery in a number of areas. The French have created entire electrical transmission systems as well as metros that they export. These are areas that the United States should be able to compete in because of our large internal market.

Rather than abandon our industrial base, we should try to think "outside of the box." One way is to move toward the idea of combining the customization of the supply chain with the customization of the production process. Another way is to find manufacturing systems that reduce global warming and export these to various parts of the world that are resisting the idea of carbon footprints. Again, let us turn a problem into an opportunity for the United States to restore its innovative edge. In Chapters 5 and 6 we return to this issue of how to restore our manufacturing base in the discussion of of extension manufacturing services and public-private cooperation.

THE EVOLUTIONARY MODEL AND
INSTITUTIONAL ANALYSIS

The Introduction ended with the observation that we needed a new policy model that emphasizes evolution. The same argument could be made about institutional analysis and especially relative to the problem of innovation.

Institutional analysis is a vast improvement over the idea that all societies were becoming like the United States, a view typical of the social sciences in the 1960s through the 1980s. Rather than adhere to the belief in a single model of society, institutional analysis argues that each society has different cultural patterns that have been slowly formed over the historical past, even ones that go back centuries. Therefore, the testing of the idea innovation network model in four different European countries, when the model was developed in the United States, was important because its generality was being tested. Although the model recognizes that societies are different, it tends to treat different sectors within the society as the same because of institutional patterns at the societal level.[51]

A particular perspective within institutional analysis highly relevant to the study of innovation is called the national systems of innovation.[52] But consistent with the previous observation, it argues that there are typical patterns of innovation that one finds in many sectors of the society. For example, the argument has been made that the United States has the pattern of radical innovation, whereas Japan and Europe have the pattern of incremental innovation.[53] The national systems of innovation perspective does recognize that some societies will be more innovative in particular sectors; thus there is some recognition of sector differences, and the explanation for which particular group of sectors lies primarily in the institutional patterns characteristic of the society.[54]

Both institutional analysis in general and the national systems of innovation in particular are missing a model of structural change. The specific evolutionary model advanced in this chapter solves this problem, but it is a model that focuses on the institutions of science and technology, whereas typically institutional analysis emphasizes economic and political institutional patterns. Institutional analysis does have a powerful perspective for explaining why societies do *not* respond to structural changes, and that is the idea of path dependency, which has already been introduced in our discussion.[55] For example, in this book, the argument is that evolution in science and technology has produced failed evolution because of path dependency. Thus this evolutionary perspective fits within institutional analysis comfortably and allows for more subtle analysis on a sector by sector basis. Furthermore, the evolutionary model advanced in this chapter can be usefully combined with the national systems of innovation thinking and another model called varieties of capitalism. One can argue that the failures in evolution in the Anglo-Saxon countries such as the United States tend to be failures in cooperation because of the emphasis on competition, whereas the failures in the coordinated market economies typical in most of Europe are more likely to be failures in the creation of small high-tech companies based on risk aversion. Evidence for these assertions is presented throughout the book.

But while I argue that a sector analysis is critical for all the reasons that have been given in this chapter and the Introduction, the societal perspectives on institutional rules, and particularly those that block change within institutional analysis, are still relevant to the kinds of analyses that need to be completed by policy makers when assessing failure in evolution. At various points, I have introduced these institutional patterns to explain why the United States has failed to change, most notably in the analysis of the dominant business model in the previous chapter. This produces a cognitive path dependency, an important line of reasoning within institutional analysis. This is paralleled at the social psychological level with the ideas of single-loop thinking, or path dependency, and double-loop thinking, or the capacity to change one's business model.[56] In Chapters 4 and 5 different examples from other societies will be provided.

Another contribution of this model to institutional analysis is that unlike most of institutional analysis that focuses on the level of society and its institutions, the eight obstacles are at four distinctive levels of analysis: research team, organization, sector, and society. Beyond this but with less consistency, the globalization of both the idea innovation network and the supply chain is considered as well. In particular, because this evolutionary model starts at the level of the sector, it can more easily unite analysis at these four different analytical levels. In particular, this four-level framework broadens considerably the usual analysis of why managers and policy makers do not change their routines or path dependency. At various points in this book, I emphasize that the CEOs of business organizations and the directors of the national research laboratories rather than policy makers make many of the bad decisions.

Although there is not a model of structural change in institutional analysis, the idea of institutional entrepreneurs, transformational leaders that attempt to change the institutional patterns of society, is one important source of change.[57] This idea is extremely relevant, and examples of transformational teams, my term for institutional entrepreneurs, are provided. In conclusion, the evolutionary model adds a great deal to institutional analysis and is at the same time richly informed by it.

2 BROADENING THE VISION OF RESEARCH TEAMS
Step Two in Restoring the Innovative Edge

THE INTRODUCTION advocated radical innovation as the remedy for restoring the innovative edge, while the previous chapter presented strategic opportunities where this remedy would appear to have the largest payoff in protecting American employment. But how does one create radical innovation? According to the management of innovation literature, the engine for scientific breakthroughs and major technological advances is the complex research team, meaning the diversity of perspectives.[1] A good corporate example of a firm that has given us many radical innovations, including GORE-TEX (waterproof material), Elixir (top-selling guitar string), Glide (dental floss), and medical products such as heart patches and synthetic blood vessels, is W. L. Gore & Associates. The secret of the company's successes is the self-managed teams that regroup in response to changing needs. All employees reserve 10 percent of their time to pursue speculative new ideas and may work on them for years before a product is brought to management.[2]

The advantages of multiple disciplinary teams are demonstrated in a National Science Foundation (NSF) program designed to foster interdisciplinary collaborations to build better information technology. In an evaluation of the entire program of sixty-two projects, the percentage of innovation was identified in one of the following kinds of innovation:

1. started a new field (58 percent)
2. developed a new methodology (66 percent)
3. was recognized with an award for their accomplishments (19 percent)
4. created new software (71 percent)

5. created new hardware (13 percent)
6. submitted a patent application (15 percent)[3]

Independent of this NSF study of multiple-discipline projects, a similar idea is found in the health sciences where the focus is placed on the advantages of both multidisciplinary and transdisciplinary research teams, that is, teams interacting across their diverse expertise areas either sequentially or at the same time.[4] The importance of diverse teams for creative problem solving has also been demonstrated in the engineering or project management literature.[5] In other words, in four distinctive literatures there is a common prescription for stimulating radical innovations.

In Chapter 1, I made the argument that restoring the innovative edge requires radical innovation. But this prescription has to recognize that what many academics and researchers in the national public research laboratories emphasize is *normal science,* the pursuit of incremental innovations. In industrial research laboratories, sometimes the objective is a radical breakthrough, most typically in the pharmaceutical and semiconductor industries—again, an important variation among sectors. But many researchers in industry focus on minor modifications of the products that the firm sells. Therefore, one obstacle to restoring the innovative edge is the failure to emphasize high-risk research. Both the academy report and the several books discussing the innovation crisis, as indicated in the Introduction, have agreed upon this as a stumbling block.

As further evidence of this problem, a recent article argues that the war on cancer had largely failed because high-risk research was not being funded.[6] Instead, scientists suggested that grants had become kind of a jobs program keeping laboratories working on what I have called normal science. In some respects, the critique about not winning the war on cancer ignores that there has been steady progress in the reduction in death rates from cancer, more in some morbidities than others, over the past 40 years, to say nothing about the realization that the disease process is now perceived as much more complex than the former single-gene therapy presupposed.[7]

One goal in this chapter is to advocate breaking these routines of normal science by increasing the *degree of radicalness* in the research objective. Advocating more high-risk research in the public or the private sector, provided that funding is available, is certainly desirable, but this still begs the question of how one broadens the vision of research team members so that they perceive more high-risk opportunities. At the team level, then, managers need to add more diversity of perspectives to accomplish this objective. The first section of this chapter provides R&D managers with a list of seven ways in which diversity can be increased. It also expands the theory about how complexity increases radical innovation (Hage, 1999).

But this simple formula for increasing the diversity of the team or teams requires recognizing the kinds of research teams. Two dimensions for describing the different research teams come from the work on research profiles conducted by Jordan (2006) in several different national research laboratories and in one industrial laboratory.[8] The first dimension is the relative emphasis on radicalness. As we have seen, this dimension varies considerably. The second dimension is the breadth of the scope of the research problem. The third dimension concerns the status of the evolutionary process described in Chapter 1, that is, how far the process has unfolded in a specific sector. In other words, research teams have to be constructed in different sizes and shapes. What the management of innovation literature has failed to recognize is that complex research teams come in different sizes and shapes or compositions of diversity.

Although the thrust of this book is to advocate scientific breakthroughs and major technological advances, the reality remains that even radical innovations vary in the degree of discontinuity. In science the degree can vary from a major finding to one that is recognized by a Nobel Prize or other signs of international recognition, and beyond to the creation of a new scientific paradigm. It must be recognized that in science, many studies replicate findings with certain slight additions to the general ideas. In technology, the degree of radicalness can vary from a minor modification to a significant improvement in the performance of the product, to the creation of a new market niche, to the creation of a new market. Obviously, I am not advocating that every research team attempt to win a Nobel Prize or that every industrial company attempt to create a new market niche. The recommendation regarding the first dimension is to disrupt the routine by simply increasing the degree of radicalness rather than advocating the extremes. However, increasing the degree of radicalness in some sectors that already have this quality as an objective poses a challenge, which is discussed later in this chapter.

The second dimension refers to how many scientific problems or technical issues are involved in the research program, that is, is the project really more like a program because of the systemic quality of the research problem being investigated? In science there is a useful distinction between Big Science (high-energy physics) and Little Science (molecular biology). In technology, there is a distinction between few-component products (cell phones) and many-component products (cars). As these few examples make clear, there are considerable differences in the relative scope of the kinds of scientific problems or technological advances to be pursued. A useful way of thinking about the scope is the size of the research effort as measured by the number of individuals or teams, budget, and more subtly, the cost and variety of measurement instruments used in the research. Even when the scope is large, the research program might be organized in different ways. It can be one very large team, as in the Manhattan Project, or it might be a number of coordinated teams, as in the Human Genome Project. Typically, in the design of complex products such as airplanes, automobiles, high-speed trains, and space shuttles, major components are assigned to separate teams within the program.

In Chapter 1, the evolution in the innovation process was outlined along with several examples. This evolutionary theory about the increasing differentiation of arenas with separate research organizations and by implication teams also has consequences for the composition of the diversity in the focal research teams. As the process of differentiation unfolds, managers have to be concerned about whether they have representation of at least some of the diversity that is outside the firm for effective communication. The third dimension is the representation of arenas.

But this process is not the only way in which the evolutionary model presented in Chapter 1 relates to the world of the research team and its objectives. The following connections can be made. First, there is a steady increase in the complexity of the research problems whether in the academic world, in public research laboratories and mission agencies, in the large industrial laboratories, or in the small high-tech companies. Increasing complexity of these problems means that their solutions require a more radical, risk-taking approach. Second, part of this evolution in the nature of the problems is the steady recognition that many, if not all, problems are embedded in a system, that is, the scope of the problem continually increases. Again, just as we can expect that the diversity of the program will change across time with the evolution in the various kinds of problems that have to be solved, so we can expect the size of the budget over three to five or more years to be the best indicator of scope.

But whereas the radicalness and scope of the research problems lead to more diverse and larger teams, the process of differentiation reduces the size and diversity. Managers thus have a challenge to balance these conflicting trends and at the same time attempt to overcome the obstacle of continuing along the same path of research objectives. In other words, even the remedy has to be continuously reevaluated.

One might wonder why a manager would reduce the number of arenas represented in a diverse research team given the imperative of maintaining communication links of one kind or another with the differentiated arenas, as was indicated in Chapter 1. The reason why communication links between several diverse teams replace one very large team is that team membership absorbs more time than maintaining communication links. Furthermore, as is argued earlier in this section, the problem of communication *given too many paradigms* becomes more and more difficult to overcome. Admittedly, special mechanisms are needed to ensure effective communication between the differentiated diverse teams in the distinct arenas of the idea innovation network (see Chapter 5).

Unfortunately, one of the recurring themes of this book is that there are no simple solutions. Increasing the radicalness of the research objective, broadening the scope of the problem, and adjusting the representation of arenas within the team contain dilemmas, ones that are above and beyond the basic issue of communication between different paradigms.[9] Too often, management consultants suggest solutions without indicating some of the economic or social costs involved. The construction of diverse research teams does have certain social costs attached to each dimension. These costs are best expressed as dilemmas, or potential blockages, that have a common theme of choosing between the following:

- individual scientific freedom or team objectives
- program autonomy or organizational coordination
- organizational autonomy or interorganizational coordination

Resolving these dilemmas is more difficult than it might appear at first glance.

The remaining sections of this chapter discuss the dilemmas attached to each dimension of diversity. In each section, some suggestions about solutions to these dilemmas are provided. As is well known, constructing diverse research teams with one of the strategic opportunities discussed in Chapter 1 may require a specific expertise, which can only be obtained outside the university, the national research laboratory, the industrial firm, and especially, the small high-tech company.

CONSTRUCTING DIVERSE RESEARCH TEAMS

What makes a team diverse? How would a manager begin to construct such a team? The definition is combining two or more ways of thinking about the specific scientific or technological research problem at hand in at least three individuals who are working toward the same goal and who interact at least several hours a week about the common goal.[10] The funding may come from more than one source, and multiple sources of funding do appear to be related to more innovation.

A pharmacologist, a health psychologist, and a neuroscientist conducting a collaborative study on the interrelations among nicotine consumption, brain chemistry, caloric intake, and physical activity levels is an example of interdisciplinarity in medical research.[11] In graduate and professional schools, students are taught to think in certain ways, sometimes referred to as scientific paradigms or schools of thought. The co-founders of Modumetal, described in Chapter 1, are Christina Lomasney and John Whitaker, who constitute another example of a diverse team.[12] She has a diverse educational background, having trained as a physicist in college, and then having held various managerial positions involving nuclear research, as well as having been a project engineer at Boeing. He is doing doctoral studies in chemical engineering on the specialty of electrochemical printing. Beyond this diversity of the co-founders is the diversity of the research team that has been recruited, which includes the following disciplines and specialties: mechanical engineering, chemistry, metallurgy, and structural physics.

Once the diverse research team is created by combining two or more paradigms, then, by definition, the differences in thinking attached to each paradigm act as a barrier to communication. Because this barrier is a problem in all diverse research teams regardless of their strategic objectives, the solutions to this problem become the focus of Chapter 3, where three sets of practices are recommended as a way of encouraging interdisciplinary communication. In this chapter, however, our concern is with the varieties of ways in which the diversity within research teams can be increased so that there is a broader vision.

The great advantage of scientific paradigms is that they can be taught and learned; they frame what are important research problems; and they provide ways of thinking about these puzzles. But this great strength becomes a great weakness relative to developing radical product and process innovations because, while providing focus, it narrows possibilities. The blinders of paradigms can also influence some of the most creative of individuals. For example, Galileo refused to accept the existence of comets because they did not move in circles like the planets. And Einstein was highly resistant to the new paradigm of quantum mechanics.[13] Therefore, to think "outside the box of the paradigm," one needs to break the routine patterns of thought in some way. Sparks of creativity occur when different paradigms are encouraged to interact. Some examples of these "sparks" are the radical innovations of nanotechnology, bioinformatics, and neuroscience.[14] To achieve this type of interaction does not mean necessarily to deny the importance of the paradigm as an initial starting point, but it does imply that one has to think about the possibilities that are not exactly covered or handled well within the existing paradigm and even the necessity of creating a new paradigm. For example, new paradigms of oceanographic science and cognitive science emerged from interdisciplinary cooperation.[15] Chapter 4 highlights the creation of the new specialty of biomedicine by combining basic biological scientists with medical researchers. Thus the easiest remedy for overcoming normal science, encouraging creative sparks, and thinking "outside of the box" is to combine different ways of thinking, or diverse paradigms, in teams. Indeed, the definition of a complex research team is precisely that the team has "two or more ways of thinking."

The second advantage of a diverse research team, the one that explains why in many cases it is superior to the reasoning power of a single individual, no matter how creative, is that as the different paradigms confront each other, they expose weaknesses in the ideas attached to each.[16] This leads to the correction of mistakes and the selection of which ideas are the best. It is this potential to pierce through misconceptions that is especially valuable in developing scientific breakthroughs and technological advances. The same is true during the creative process as scientists and engineers develop solutions to some problem. As I have already observed, even good ideas are half-wrong, and the advantage of the team is sorting out which parts are good, which are not, and, more critically, improving upon the half-wrong ideas. But this process assumes that complex research teams have effective internal communication, which is anything but certain and therefore deserves a whole chapter.

Managers can assess how diverse their research teams are by examining the list in Table 2.1 and constructing their teams accordingly. Although the list appears to emphasize scientific examples, in part because this is so critical for restoring the innovative edge, the same logic applies to engineering specialties or other ways of categorizing departments within large public research

Table 2.1. Sources of diversity within the research team

Sources of diversity listed by extent of differences in thinking

- Separate theoretical and empirical skills in a research team (e.g., methodologist, technician, statistician)
- Diverse schools of thought or even paradigms within the same specialty (e.g., the French and German paradigms of immunology)
- Different specialties within a discipline (e.g., biochemistry, biophysics, genetics within biology)
- Distinct disciplines (e.g., physics, chemistry, and biology)
- Various arenas of research (e.g., basic research, applied research, product development)
- Dissimilar organizational cultures (e.g., DuPont, Dow Chemical, Monsanto within the same general sector, in this case chemicals)
- Dissimilar national cultures (e.g., American, French, Russian)

organizations or industrial research laboratories within firms. When one shifts to the world of technology, frequently the operative terminology is capabilities. For example, General Electric (GE) identified twenty-two capabilities that it needed to create a broad-scope program of research on coal gasification. The internal analysis of the managers of GE identified that twelve of these capabilities were within the company, six were held by their partners, but four were completely outside their interorganizational network.[17]

The list in Table 2.1 is, of course, a discrete example. The list would be quite different for each specific program of research, and therefore I cannot provide in this chapter a generic list of capabilities. But the logic of constructing a diverse research team remains the same, namely, to combine capabilities so that the new product or major modification advances the technology radically beyond its present state.

Connecting this idea to the discussion in Chapter 1 about attempting to solve global problems and address national concerns, these objectives can be translated into sets of capabilities. For example, to reduce energy consumption in the manufacturing of a specific product would mean adding certain capabilities to the research team. As more and more desired performances are added to the product, then more and more capabilities are needed in the research team.

Counting the number of different paradigms or capabilities in the team is not a static issue. Diversity or complexity in the skill sets and specialties of the researchers and technicians can change across time. As research findings emerge, managers begin to recognize the need for still other kinds of equipment or other kinds of expertise, that is, new knowledge areas have to be added to the team. Nor can the manager only worry about this at one time point. Both as an advantage and as a disadvantage, the specific disciplines or specialties that compose the capabilities in the research project usually alter over time as new problems present themselves that require different subspecialties or competencies. In other words, even in the course of a specific research project, as understanding grows, there is an evolution in the nature of the problems that have to be solved. Again, the fluidity of the team composition has been missing in the discussions about the advantages of complex teams. In health care research, as we have seen, a distinction is made between multidisciplinary and interdisciplinary; in the former, the contributions of the different disciplines are made sequentially, whereas in the latter, they are made jointly by working together.[18] Our main focus here is on what they would call interdisciplinary and transdisciplinary teams in which the research work is completed jointly with varying degrees of intellectual integration of paradigms. The obvious point that teams need to change across time is illustrated in our research on Sandia National Laboratories.[19] The research projects and programs each year typically added new scientific and engineering specialties and, in some cases, dropped others. In other words, the construction of a complex research team should recognize the temporal dimension as well.

In addition to using the list in Table 2.1 to assess diversity in research teams, managers can use the list to assess the level of barriers to communication within their research teams. As one moves down the list, the differences in thinking patterns, skill sets, assumptions, and methods become greater and greater, making effective communication more difficult. The distance between subspecialties within a discipline is usually much less than between disciplines, which becomes critical for managers as they assess the likelihood of barriers to communication developing within their diverse teams. Obviously, the greater the number of different paradigms and the greater the distance between them, the greater the effort that must be exerted to overcome the communication barriers imposed by these differences.

The first kind of diversity to build into a research team is dependent upon the different skills needed for research. The importance of skill variety is illustrated in the study of Thomas Hunt Morgan's research team at Columbia University that discovered some of the basic genetic principles of inheritance.[20] This research project did the pioneering work on genetics working with the fruit fly. Although Morgan received the Nobel Prize, the reality is that his young team included a diverse set of skills, among them a program manager, a designer of experiments, a manager of the stock of fruit flies, and a theorist who actually conducted the research experiments.

Many studies of science indicate that a team needs both an idea generator and a critic who indicates which ideas should be selected and, most critically, how they should be modified. What is important about this distinction is the recognition that even the role of theorist or idea person has several varieties, reflecting different kinds of creativity. Also, what this suggests is that many of the brainstorming techniques that are used are only a very small first step. The real step is to begin to understand how the initial insight should be modified and refined. In my research on the Pasteur Institute, which is reported in Chapter 4, I found that Pasteur would generate many ideas, but both Emile Duclaux and Emile Roux were good critics that refined his ideas in important ways. This process goes back to one of the great advantages of a team approach: A good idea depends upon useful criticisms to improve it.

Closely akin to the idea of different theoretical and empirical skills is the nature of certain social roles in the research team. One key role in a research team is that of the technological gatekeeper, the individual who scans the research environment to discover the latest research findings.[21] With the new stress on the importance of *open innovation*, that is, finding promising ideas in the scientific and technological environment, the technological gatekeeper has an even more critical role in the research team now than when first identified some thirty-five years ago. Other social roles have been identified in the R&D management literature; besides the idea generator and the gatekeeper, they include the entrepreneur and the program manager. What these various roles suggest is that managers need to think about the skill set that is necessary within the team and recognize that different individuals are better at one or another of these roles as well as having different skill sets. Furthermore, which set of social roles one thinks about will vary from the kinds of complex research teams found in science and those more concerned with innovation in firms. An example of how a typology of social roles can be used to create a diverse team and one that is more likely to have high rates of communication is provided in an extended case study in Chapter 3.

The second source of diversity comes from the recruitment of researchers from different schools of thought or even paradigmatic representations within the same discipline. For example, within economics, combining a representative of the neoclassical model of thinking with an evolutionary economist would create a complex economic team. Or within biology, in much the same vein, combining a Lamarckian with a Darwinian. Distinct paradigms within an occupational specialty are not always well developed, so that the more useful recruiting principle for managers is to select people trained at different schools. Usually, they will think differently about the same problem that needs to be solved. The importance of not favoring one particular intellectual source for a discipline or subdiscipline is illustrated in the study of the Pasteur Institute during its early days.[22] As indicated in Chapter 3, one mistake made by many managers is to select only individuals from the most prestigious universities and not cast a broad enough net to find creative individuals at the less prestigious places.

Some of the research studies in the management of innovation have measured how people do, in fact, think differently. One study demonstrates that various functional departments within the same firm have different funds of knowledge or expertise and systems of meaning that create barriers.[23] This finding documents again the basic insight about the conflicts in the *paradigms* of research, sales, and production within the same firm.[24]

As a policy objective, Stokes (1997) has made much of the idea of combining basic and applied research, which he calls Pasteur's Quadrant. In fact, the in-depth study of the Pasteur Institute during its first thirty years indicates that many of the radical innovations were made by teams composed of a biologist (basic) and a medical researcher (applied). These two selections reflected Pasteur's concern with creating a kind of physics of biology, even though the specific focus was on bacteriology, and then combining this with medical research.[25] But in addition, many of these teams also engaged in research on how to improve the manufacturing and the quality of the vaccines and serums that were sold to help support the research efforts of the Pasteur Institute. Thus, in effect, the diverse teams generally involved four, if not more, of the six arenas of the idea innovation network (see Chapter 4). Therefore, Stokes's understanding of how diverse the teams actually were is limited.

Finally, the last way of creating diversity is the integration of diverse cultures within the same research team. As we shall see in Chapter 4, of the eight and one-half radical innovations in biomedicine at the Pasteur Institute during the first thirty years, four were the contributions of foreign scientists who permanently migrated to France and worked in the institute. In particular, the team efforts of Elie Metchnikoff, a Russian zoologist, and Roux, a French physician, and their different ways of thinking—the former speculative and the latter critical—left an indelible mark on the minds of the various students who worked with them. In other words, two kinds of diversity were built into the same complex research team. Nor is this an unusual finding. Many of the winners of major scientific prizes in the United States have been immigrants. Combining unlike national cultures is probably one of the more profound ways in which to measure exposure to different ways of thinking, because the national systems of education, especially at the level of the university and the medical school, have different kinds of programs and views about what constitutes knowledge, and thus distinctive ways of thinking.

Given how much variation can exist within each of the categories listed in Table 2.1, a practical manager might want to know how to obtain a good sense of exactly the degree of diversity within the research team that he manages. One procedure for assessing this diversity would be to ask each individual which journals or sources of technical information he or she uses. This quickly provides a sense of the range and depth of the set of capabilities with the research team as well as particular opportunities for communication if several individuals read the same journal or rely upon the same source of technical information.[26]

DEGREES OF DIVERSITY

First Degree: Increasing the Radicalness of the Objectives

In the Introduction, a large number of examples of radically new products and processes were provided. But what would be a scientific breakthrough or radical advance in science? This question is much more difficult to quantify, and yet for managers it is important to define this as well. Scientific breakthroughs are discerned relative to the context of the knowledge pool, that is, how much is known. The breakthrough represents a significant advance in understanding. These advances can occur in a variety of ways in science, one reason why it is difficult to quantify the extent of the advance. Recognizing the radicalness of a scientific advance may mean finding a solution to a central problem, discovering a phenomenon that was unknown previously, obtaining research findings that upset existing theory, and obviously and most critically, developing a new scientific paradigm. Sometimes the radical advance occurs when a central problem is solved, such as the identification of the structure of DNA.[27] A new biomedical specialty, molecular biology, was created with the discovery of DNA and RNA. Sometimes it happens when a major discovery is made, such as the observation of the first black hole in 1971 (Cygnus X-1) in radio astronomy. In archeology, we have the example of the discovery of skeletal remains in the Americas estimated to be over one thousand years older than when the first settlement in the Americas was theorized, and in paleontology, the discovery in China of fossils of dinosaurs with feathers, which had a similar impact.

The importance of diverse teams is illustrated in all the above discoveries. The creation of molecular biology is an exemplar of combining paradigms in the same research team to increase the probability of a scientific breakthrough. For example, the discovery of DNA was the work of James Watson (biology) and Francis Crick (physics), and that of RNA was the work of Jacques Monod (enzyme research in biology), François Jacob (medicine), and André Lwoff (phage research in biology).[28] And other research teams performed the important work in crystallography that facilitated the work of Watson and Crick, and various research teams contributed to the work that helped Monod, Jacob, and Lwoff.[29]

One simple way of measuring a radical breakthrough, although admittedly only after the fact, is the awarding of some major prize for the discovery. With a team of others, I measured approximately three hundred major breakthroughs in biomedicine and where they occurred over the course of a century on the basis of a list of different kinds of prizes.[30] Again, diverse teams made most of these discoveries.[31] A good example from an in-depth study of the Pasteur Institute reported in Chapter 4 is the treatment for diphtheria. Roux's work unfolded in two periods, both within complex research teams.[32] First, Roux worked with Alexandre Yersin, his research assistant, and a Swiss surgeon, to discover the toxins that caused paralysis and recognized that they were soluble. Later, after a bout of illness, Roux turned toward an effective treatment for diphtheria, working with Louis Martin, a clinical physician, and Edmond Nocard, a veterinarian. The researchers were able to develop an effective serum from horse blood, which they tested on three hundred patients to prove its effectiveness. Roux won the Copley Medal for this work.

What about applied science or product development? And how can one measure the extent of applied scientific breakthroughs, especially if there is little commercial value to the breakthrough? In our work at a national research laboratory, we identified several examples of small projects with diverse research teams that achieved radical breakthroughs.[33] One especially interesting project was the design of a catalytic converter for diesel engines that reduced particulates by 400 percent. Another project focused on the development of semiautonomous, modular robotic control technology that decreased the necessity of human control and presence in threatening situations. Still a third example of a radical advance that led to a wide variety of different applications was the development of a simple dipstick that could do instant laboratory tests. One particular application of this wide-ranging technology was the development of a dipstick that could test for lead in water for underdeveloped countries. The commercial value of the new catalytic converter can easily be computed, but the two other technologies demonstrate that their value is expressed more in the lives they save rather than in their commercial value.

The Center for Satellite Applications and Research (STAR) of the National Oceanographic and Atmospheric Administration (NOAA) conducts very different kinds of applied science and technological research projects than the given examples. The STAR division is concerned with developing products from the information sent by satellites launched by the National Aeronautics and Space Administration (NASA). Many of these products provide information that can be used to improve the weather prediction models of NOAA. One team developed a radically new product that reports smoke from fires around the world and has been used to report fires in the Amazon as well as in the western part of the United States. The early reporting of fires in remote areas can result in enormous savings relative to the cost of destroyed forests, to say nothing about the potential saving of lives as well. Another team in a cooperative research program at the University of Wisconsin was achieving a threefold increase over the current compression rates of data streams sent by the satellites in preparation for the next generation of satellites that would be able to send much more data with which to do weather prediction. The value of this product is represented by the value of various improvements in weather prediction and other uses of the satellite information that STAR processes. Again, diverse research teams achieved these applied science breakthroughs.

Relative to radical technological advances in industrial products, at various points we have already mentioned the Apple iPod, which is discussed at length in Chapter 5, Hewlett-Packard's radical advances in new kinds of printers as well as in the way in which they are commercialized, and Modumetal's solution to the third industrial divide that can provide radical new forms of body armor. Chapter 6 has a number of other examples from the Advanced Technology Program funded by the National Institute of Standards and Technology as well.

Assessing the extent of a scientific breakthrough after the fact and by implication determining how diverse the number of paradigms has to be within a scientific research team, paradoxically, is much easier for radical technological advances. Counting the number of capabilities needed is defined by the objectives desired in the performances of the product being developed. The more performances that need improvement, the more capabilities.

In our research at a national research laboratory, we have found that many research projects had only six to ten people, but in some cases these individuals represented six or more departments, whereas in other instances the amount of diversity was much less.[34] We counted the departments because separate disciplines were housed in disparate departments. Thus, even in relatively small research projects, there can be a considerable variation in the degree of diversity in the composition of the researchers and technicians.

One important and ignored source of diversity in the research team is the variety of equipment that technicians need to use in the research project. This aspect of science does not receive the attention it deserves. The addition of different occupational specialties usually includes different kinds of equipment and technicians to operate the machines and maintain them. In constructing diverse teams, it is important to remember the critical role that technicians and equipment play.

Managers of complex research teams are confronted with both visible and nonvisible dilemmas. The first dilemma occurs in the selection of the team members. They may be tempted to select individuals who have worked together before with the idea of reducing communication problems. But doing so also creates the risk of groupthink, which reduces the possibility of a radical advance.[35] By reducing the differences in thinking, the managers are reducing the creative dialectic as well (see Figure 3.1). On the other hand, selecting individuals who have had a history of intellectual conflict will prevent any possibility of overcoming the barriers to communication created by different paradigms because of the considerable drop in communication. The resolution of this dilemma is selecting independent thinkers who are capable of working together. An extended example of how to resolve this dilemma in a new research unit created at a national research laboratory is provided in Chapter 3.

Another dilemma, or blockage, is present in the reward system, one of those nonvisible components of the context that can create considerable difficulties for managers of research projects attempting to integrate their team members. If the reward system is based on individual performance rather than team performance, it creates the conditions for a lack of cooperation within the team. On the other hand, just rewarding team performance can create other kinds of problems; that is, some individuals will not contribute as much effort as others to the team's performance.[36] Again, the dilemma for managers is striking a balance between recognizing individual contributions and team performance, which is not easy to achieve.

The reward system is not the only structural condition that might make the manager's job of integrating the diverse ways of thinking in a research team more difficult.[37] Another is the physical location of the individual members. It might seem obvious that the team members should all be in the same location so that they can easily interact and communicate. But this arrangement ignores the way in which subspecialties and disciplines are housed in large universities, public research organizations, and industrial research laboratories. Typically, they are located in separate and homogeneous departments, in which a subculture forms that leads to groupthink and also stovepipes, as was mentioned in the Introduction. This same problem exists when research organizations are structured around research programs. A new research program connecting two programs might result in the individual researchers being located in separate buildings. Unless they are colocated, distance is a special obstacle. Nor are these the only blockages that create difficulties for managers.

The central task for a manager of a diverse research team is reconciling the goals of the individual team members with the team goals. This task relates to the dilemma noted at the beginning of the chapter about academic freedom. Individuals like to explore their own new ideas and not necessarily pursue a larger objective. Therefore, the real obstacle in many small research teams is the vision of the project leader. To make him or her perceive the advantages of a larger vision that would increase the radicalness of the research objectives requires resourceful managers, especially in universities where these problems are the greatest.

An intensive study of creativity among researchers across many disciplines in different contexts—academic, government, industry—indicated that when individual goals are reconciled with collective goals, whether the team's or the larger organization's, more productivity (measured by the number of papers or patents) and creativity (measured by peer review of the papers and technical reports) occurred with shared decision making between the scientists and the management.[38] In other words, there is autonomy but within general strategic guidelines. Furthermore, this is an excellent illustration of how combining diverse goals can create a positive stimulus for ideas, the main theme of the previous section. It may seem strange, but giving complete scientific freedom to researchers is not necessarily in their best interests *if the goal is radical innovation.* But reconciling the individual and team goals requires not only good negotiation skills on the part of the manager but also some creativity in seeing possible ways to accomplish reconciliation.

A good example of how this reconciliation might be accomplished comes from a study of biotech firms. In the more innovative firms contrasted with those that were less innovative, management retained control over general strategy, but they gave the researchers autonomy over the decision making affecting their research. To quote from one of the managers: "Top management and I are constantly specifying the ultimate goals to be attained, but we are careful to allow our scientists creative ways of achieving those goals."[39] Another and more extended example of how to resolve this dilemma between strategic objectives and researcher goals is provided in Chapter 3.

In summary, managers of research projects considerably increase the probability of making a radical advance in either science or technology if they can ensure that the individual goals and team goals are complementary, if the reward structure reinforces this balance, and if the members of the team are co-located. It is worth repeating that the issue is not striving for the great breakthrough but instead broadening the vision somewhat beyond the present range of the research team's objectives. Similarly, this broadening of vision does not require the addition of a great deal of diversity but simply more than is currently present.

Second Degree: Increasing the Scope of the Research Problem

The scope of a research problem is usually inherent in the nature of the scientific or technological problem to be solved. One simple way of determining its scope is counting the number of parameters in the scientific or technological system involved.[40] Some of the scientific disciplines have a systemic quality, that is, a large number of variables that have to be considered at the same time. Examples include astronomy, oceanography, and meteorology. Hurricane forecasting involves very complex prediction models that, in turn, rely upon a very extensive system of measurement, including planes, ships, and, most critically, satellites. Typically, the analysis of the large amount data collected by the National Weather Service (NWS) is constantly exceeding the capacity of its large computers.

Broad-scope innovation requires the creation of a research program as distinct from a research project. With this in mind, we define a *research program* as a group of research projects that all have the same goal, namely, advancing either knowledge or technology in a particular discipline or problematic arena, *and that are coordinated as well*. When such a focus is necessary over a long period of time and when its responsibility lies with the government, the common response is to create a mission agency. One of the earliest examples of a large-scale program of research is the famous Manhattan Project that produced the first atomic weapon in the 1940s. Later the mission agencies of Los Alamos and Sandia National Laboratories were created, the former to continue research on atomic energy and the latter to focus on the problem of stockpiling and safeguarding atomic weapons. In the 1950s, President Eisenhower signed the Act of 1958 which created NASA in response to the Soviet Union launching *Sputnik*. NOAA was established in 1970 to unify and coordinate the government's research efforts on various aspects of the global environmental system, including the NWS. Altogether, the scope of NOAA's mandate necessitates expensive and specialized equipment and teams to collect the relevant data, including satellites, ships, buoys, and planes. Indeed, one indication of a broad-scope problem is the necessity of collecting data on a large scale. The most recent example of a new mission agency with a broad-scope research program is, of course, the Department of Homeland Security (DHS).

Some research laboratories that have many small research projects can also conduct major research programs of broad scope. A good example is the National Institutes of Health (NIH), which conduct some major research programs in the National Cancer Institute (NCI), the National Institute of Drug Abuse, and the Clinical and Translational Science Centers, among others.

Broad-scope research programs also have to be created when the scientific or technological problem is difficult or requires research under some extreme conditions. Consider the case of conducting research on flora and fauna at the bottom of the ocean or of studying black holes at the other end of the universe. Natural systems that are at the extremes of scale, from the very small (subatomic) to the very large (outer space), usually necessitate expensive equipment, numerous technical personnel, and many researchers. The extremes in "Big Science" are reflected in the cost of research on particle physics found at the European Organization for Nuclear Research (CERN), to say nothing about the building of a space vehicle to go to Mars, estimated at $20 billion. Each increase in accelerator beam energy means an additional cost and even a different kind of machine as, for example, cyclotron, betatron, proton linac, proton synchrotron, and superconducting proton linac (SPS) collider.[41] Indeed, as the example from particle physics suggests, it is important to note that a major cost item can be the research equipment, such as particle accelerators. Indeed, high costs are often what distinguishes Big Science, that is, the kind of large-scale research that can no longer be conducted in the laboratory. Furthermore, the increases in the size and cost of the equipment to conduct the scientific experiments illustrate the growing complexity of scientific research and thus the need for more complex research teams.

If the research problem can be divided into smaller research projects, then, obviously, this division is the preferred choice. Many topics examined in health research are capable of this kind of compartmentalization. But if one wants to study the system in its entirety, then the research program, or even research organization, becomes the more appropriate management solution.

Parallels of large-scope projects exist in technological research. Products can be differentiated on the basis of the variety and size of their components. In industry, large-scope research programs include the creation of hybrid automobiles or new generations of airplanes that consume less gasoline, such as the Dreamliner. With a 33 percent increase in fuel efficiency, it provides an example of a radical advance in technology. The innovation of high-speed trains in France required teams that worked on brakes, others that worked on the design of the engines, still others that concentrated on the construction of the new kinds of tracks, and so on. Even the ticket ordering process was redesigned. All of these teams had to be coordinated. The failures of coordination across many teams have recently been illustrated in the difficulties that Boeing encountered in delivering the Dreamliner on time.

Another way of measuring whether a breakthrough has occurred is assessing the difficulty of the technological problem, a parallel to the problem of studying certain scientific problems under extreme conditions. The building of a racing car reflects an extreme condition that requires all kinds of specialized metals, motors, and tires, to say nothing about the high-tech safety devices. The development of military equipment to function under extreme conditions, such as those of heat, sand, and water, is another illustration.

How do the evolutionary patterns described in Chapter 1 impact on the scope of the research team? The movement toward customization, combined with increasing demands for technological sophistication, better design in the products, and, in particular, higher quality, necessitate that a considerable array of capabilities be included in the research teams. Simply determining the advance desired in certain functions makes apparent how radical the objective is and what kinds of research expertise are needed. If the product is a complex one or has multiple functionalities, then again the scope is broadened accordingly.

Another implication of the evolutionary patterns described in Chapter 1 is the emphasis on national and global concerns. I stressed the problem of carbon footprints because it applies to the manufacturing of all products, although more to some than to others. Attempting to reduce energy consumption in the manufacturing and operation of the product being sold probably implies a broad-scope research team to redesign the product and another to redesign the manufacturing process.

But there is an important qualification to this continual increase in the size of the research team. Just as the evolutionary processes in how innovation is produced lead to a potential reduction in the size of the team—because certain competencies are now located in other research organizations or firms— the presence of a supply chain, particularly for complex and multifunctional products, suggests that there will be a division of labor in the research on the product. As is well known, cars, trains, planes, and shuttles are products with multiple components, some of which are supplied to the coordinating company or mission agency that assembles the components into the final product. In Chapter 5 is an extended discussion of the supply chains for cars (highly complex products with moderate technological sophistication) and semiconductors (low complexity but with very high technological sophistication), again illustrating the differences between sectors. Sometimes one component can be changed in a complex product without major implications for the other components; for example, the catalytic converter was added to cars without calling for too many changes in the motor. More frequently, attempting to make a major technological advance such as greater fuel efficiency necessitates innovation in many components.

Radical advances in broad-scope projects typically entail larger budgets. The simple movement of increasing budget sizes roughly corresponds to the movement from a project to a program to a research organization (small high-tech firm) to a mission agency (NASA) or firm that has a large research budget (as in the semiconductor industry, automobile industry, and aircraft industry). Furthermore, as is obvious, as the costs increase—the movement from $10 million to $100 million to $1 billon to $10 billion—there is a need for significant changes in the organization and management of the research teams.[42]

At the extreme is the cost of a space shuttle, with a price tag of $20 billion and a duration of ten to twenty years of research on developing the new technologies required.[43] From this characteristic flow many of the dilemmas that emerge in the coordination of broad-scope research programs, our next topic.

Research programs of a large size require a significant investment in managerial control and coordination. Applying control and coordination rules creates tensions. The dominant source is the struggle for project or team autonomy. Although the problem of project autonomy also exists in the small-scope research teams—I could have discussed it in that context—the dilemma of project autonomy becomes much greater when multiple research teams in a program need to be coordinated, and even more so in the large mission agencies such as DHS, NASA, and NOAA and in the supply chains for complex products such as in the designing of a new car model (see Chapter 5).

The first major coordination tension associated with broad-scope research programs is the application of bureaucratic rules, the simplest but not always the most effective control mechanism. A study of engineering projects provides a vivid description of how managers of large-scale research projects attempted to bureaucratize the coordination of the disparate parts of the research program or teams.[44] But, in fact, proliferating rules or bureaucratization is what creates many of the problems within research programs. In addition to the impersonal attention to the researchers, who generally need recognition, bureaucratic control also has negative impacts on technical management in requiring time for administrative tasks that detract from time spent on research. Indeed, in a recent study of satisfaction in a large public research laboratory, the scientists and engineers complained that more of their time was spent filling out forms to ensure compliance with various regulations regarding national security (one of the unrecognized costs of 9/11), laboratory work safety, and so forth. At the opposite extreme is the case study of a research organization, an organization capable of achieving many scientific breakthroughs in a relatively short time period, which is reported in Chapter 4.

One bureaucratic rule that impacts on the creativity of the researchers is the inherent conflict between maintaining progress and schedules—an important issue in broad-scope research programs—and the freedom to pursue new directions. Although measuring and tracking progress is essential, it must be done in a way that does not force researchers to focus on shorter-term, incremental, measurable products or take too much time away from interesting links of inquiry. An important side effect of too much emphasis on schedules is the reduction in the opportunities to create a radical breakthrough. Highly innovative biotech companies stressed the importance of reasonable goals and de-emphasized deadlines. As one manager in a biotech company stated: "We have learned from the past that it is best to try to minimize the number of major projects each person is assigned to. If they have too broad a range, often they get lost. They can never do anything except step from the top of each pile to the next."[45] Indeed, this quote points to one important cause of the concern over schedules: the problem of researchers having to deal with multiple projects, some of which are in a large research program.

The second tension is the potential reduction in motivation among researchers because of the large size of the research program. As is well known, researchers are motivated as much by the recognition of their work and the intrinsic pleasure of doing their research as by extrinsic rewards.[46] Consider the hundreds of names that appear on publications associated with large research equipment such as CERN. These extremes in broad-scope programs make it difficult for individual scientists or engineers, to say nothing about the technicians that keep the machines working, to receive their due. Therefore, managers have to find ways of recognizing the contributions of each member of these very large teams. In our studies of work satisfaction among researchers in NOAA, we found that *recognition* and *respect* were two of the most important determinants of their satisfaction.[47]

Paradoxically, one attribute for motivating researchers is to ensure that they have some autonomy on a range of decisions. I say *paradoxically* because the basic dilemma in large-scale research programs is maintaining coordination, which would seem to argue against autonomy. One of the few studies to distinguish between types of research projects along lines similar to the difference between small-scope and broad-scope complex research teams found that higher project autonomy was particularly important in those projects that had broad scope. In 121 different R&D projects across thirteen research units, admittedly within the same company, providing autonomy to the project leader was associated with higher innovative performance measured in a variety of ways.[48] Providing greater autonomy would appear to be counterintuitive. But different projects within a program of research can have autonomy *provided there is a vision* and assuming that leaders can articulate this vision in discussions with the different project leaders, issues that were not studied.

This combination of some autonomy over research decisions and the importance of the vision has been mentioned in some studies of successful firms that have developed new products. In Japan, successful product development was accomplished by a balancing act between relative autonomous problem solving and the discipline of a heavyweight leader, strong top management, and an overarching product vision.[49] Similar patterns have been observed in American companies. The key is well-coordinated functional groups with commitment from top management. The successful vision insisted on concentrating on products that had superior customer value through enhanced technical performance, low cost, reliability, quality, or uniqueness.[50]

Therefore, although the growth in the size of the research program creates coordination problems, various solutions to this dilemma exist. The most important are the maintenance of a clear vision and the negotiation of project strategies within this vision. This solution necessitates a research program leader that knows how to articulate the vision in discussions with the various project leaders. An example of such a leader is provided in Chapter 3.

Third Degree: Increasing Representation of Research Arenas

An emphasis on radical innovation both as an objective and as a broad-scope scientific or technological problem increases the diversity and the size of the research team, with the important qualification about the division of research along the supply chain for complex products. Furthermore, the combination of the two objectives, that is, radical innovations in a broad-scope problem, necessitates a complex research program with multiple teams working on different scientific parameters and technological components. Typically, this combination might be done in functional teams, as in our examples of the development of high-speed trains previously cited. In contrast, the evolutionary model presented in Chapter 1 implies the opposite tendency: As various arenas are differentiated, the diversity within the team decreases as various specialties move outside the organization.

In the Introduction, the model of closed innovation, which was used by such American success stories as P&G, GE, DuPont, and Xerox PARC that gave America its reputation for being highly innovative, had more or less all six arenas in the idea innovation network within the industrial firm. But not all arenas were equally strong as the analysis of Xerox PARC shows.[51] During much of the twentieth century, the pace of product change was slower than it is now, and these firms had the luxury of developing radically new products and processes slowly and carefully, getting them right the first time. In particular, P&G has had an enviable record of successful launches of radical products creating new markets because of its emphasis on distinctive performance characteristics.[52] But in the last two decades of the twentieth century, the competitive situation changed dramatically, and speed of product development has become much more important as product life cycles have become compressed.

The movement of research arenas outside the organization, as described in Chapter 1, implies that large industrial firms cannot manage all six arenas successfully, especially in the high-tech sectors. A case study of IBM provided in Chapter 5 illustrates this point. Large firms are best advised to concentrate on three or four arenas, which is increasingly what they are in fact doing, as demonstrated in Chapter 1.

But quicker product development requires being able to transfer the hidden assumptions and methods behind each radical technical advance or scientific breakthrough rapidly; hence the strength of the connections in the idea innovation network between those arenas that are differentiated becomes critical. Besides quicker product development, another reason why innovation speed will determine whether American firms survive is the imperative that they respond quickly to the development of new products and processes in the global market when firms in other countries develop a radically new product. Given Toyota's creation of a successful hybrid, then the three American automobile companies should have responded quickly, which they did not and therefore have lost market share. The stimulus funds for the automobile industry have helped it to survive.

Given the idea innovation network model, these are the questions: How many arenas need to be represented in the diverse research team? How can one ensure tight connections between differentiated arenas? How do the answers to these questions vary by technological sector and kind of research organization in the idea innovation network? The answer to the first two questions is surprisingly the same and is discussed in the following paragraph. The answer to the third is much more complicated and discussed by examining some major kinds of research organizations and firms and how their leaders might adjust the way in which they think about the problem of constructing complex research teams that are strongly connected in a network.

Because the solution of requesting that all six arenas be represented within the same research team—the closed innovation model—is no longer a viable one, then what arenas should be retained within the research organization or the firm's teams? The answer relates to the question of how best to create a strong integration between differentiated arenas involving disparate organizations. Restoring the innovative edge requires that the differentiated research arenas in the sector have to be strongly connected in some way or else trade balances will decline, because industrial products are not being quickly developed with the latest advances in scientific and technological research that are relevant to them. The tightness of the integration is particularly critical for rapid product development and rapid responses to competitive moves made globally, and as we have seen in the Introduction, lack of integration is a failing of many American firms.

Unfortunately, the answer to this question is not to be found in the literature on innovation. Network research has not really focused on the problem of how well connected the six research arenas are because it has not considered the six arenas in the data analysis. Thus, it has missed the problem of the valley of death. Nor has it asked how strong the connection between arenas should be for effective communication.[53] Although there is a very large literature on joint ventures and networks of various kinds, none of this research focuses on the entire idea innovation network and which arrangement leads to more effective communication between arenas.[54] Nor does the model of open innovation make clear which arenas should be retained inside the company and, perhaps more critically, why.[55] As becomes apparent later in this chapter, some of the arenas of research should be retained precisely so that one can be more *open* to ideas that are outside the firm or research organization as well as outside the box. Not only should there be more research funds in diverse arenas to "pick up" on new ideas, but there is a need for a technological gatekeeper to monitor developments.[56]

The general rule for the construction of strong connections is the principle of redundancy in the number of arenas represented in each differentiated arena but redundancy of a particular kind. For two arenas to have intense interaction their research teams must overlap with the duplication of at least one arena. Taking Figure 1.3 as an example, the teams within each differentiated arena for the biotech and pharmaceutical sector should be constructed as follows:

- In the universities, the teams should include both basic and applied research arenas, even though the latter is found in small high-tech companies.
- In the biotech companies, the teams should include basic as well as applied research to connect to the complex teams in universities and

product development and also manufacturing research to connect to the teams in pharmaceutical companies.

- In pharmaceutical companies, teams should include product development as well as the other three arenas of research to connect well to the biotech companies.

The principle is that biotech companies will be better connected to the universities if both sides share one arena, if not two, in common. The same principle is applied to the linkage between biotech firms and pharmaceutical companies; they should overlap in at least product development, if not also in manufacturing research. This overlap, or redundancy, is necessary to facilitate effective communication that automatically becomes more difficult when research organizations are differentiated from each other into separate functional arenas. Not only are there communication problems between the different perspectives associated with the research in each arena but the added complication of the differences in organizational culture. As this example from biotech firms suggests, not all parts of the idea innovation network have to be connected with each other for the idea innovation network to be strongly integrated, as shown in Figure 1.3. It is only necessary that each pair of adjacent differentiated arenas be strongly integrated. Although, in general, the open innovation model means a reduction in the number of arenas, problems in the management of the arenas that are present exist and need to be discussed.

This managerial problem of how one creates strongly connected links with other differentiated areas actually has several positive aspects, despite it appearing to be a complication for managers of research teams. The first positive aspect emerges because of the increased specialization and focus, but it is not the only advantage. Another positive aspect, one of the paradoxes I mentioned in the Introduction, is that the increased specialization and focus allow for quicker development of radically new products than what occurred in the closed innovation model *provided all the differentiated arenas are properly and strongly integrated in the idea innovation network*. Strong connections to the differentiated adjacent arenas of research are the most important diversity in the research team if we are to develop radical product and process innovations rapidly and respond quickly to competitive moves outside the country. A good example of this paradox is provided in the case of IBM discussed in Chapter 5.

How effective communication between differentiated arenas is constructed varies by the technological sector and the nature of the research organizations or firms within them, because, as we have seen, the growth in knowledge has produced different differentiation patterns in the biotech and pharmaceutical sector versus the telecommunication sector. Across technical sectors, the kind of research organization that is dominant varies as well. We next consider some of the managerial dilemmas associated with each type of research organization and its prototypical complex research team.

The recommendations given for the biotech and pharmaceutical sector as an example of how to create strong connections between the differentiated arenas and, therefore, a strongly connected idea innovation network in a specific technological sector are representative of that sector only. Each of the other ten high-tech sectors, ranging from automobiles to semiconductors, would be different, not to mention the different circumstances facing the low- and the medium-tech sectors. However, the general principle in the example for increasing integration in the biotech and pharmaceutical sector provides good guidelines. At the same time, this principle generates problems that represent blockages.

This issue of how to manage the integration *between* diverse research teams across differentiated arenas is discussed at some length in Chapters 5 and 6. But in this chapter, the issue is the internal management of diverse teams, in particular the implementation of the general recommendation. To capture some of the differences that exist in concrete situations, I propose to discuss the potential blockages by kind of research organization rather than by kind of sector. I can only begin to address some of these differences across these research organizations and, implicitly, technological sectors, but I hope at least to provide a way of identifying some of the dilemmas that have to be considered.

Recognizing that this list is hardly exhaustive, I discuss the following cases in generic terms:

1. University research departments
2. Small high-tech companies
3. Large high-tech firms
4. Large public research laboratories
5. Firms, whether small or large, in the small- and medium-tech sectors

Obviously, the concrete situation for each specific firm, small high-tech company or public research laboratory, is different. It is particularly important to include the large public research laboratories, because they are typically ignored and yet they represent one of the hidden strong points in our scientific and technological innovation system. The Department of Energy (DOE) accounts for some 50 percent of RDT (basic research, applied research, and product development) spent on basic science in physics and chemistry. More visible are the many institutes in NIH, particularly NCI. One of the key advantages of discussing the large public research laboratories is that they are well situated to focus on several of the global problems that need to be solved. Furthermore, and for various reasons, both the national laboratories of DOE and many of the institutes of NIH have been moving closer to cooperating with business and thus to creating linkages between the public and the private sector (see Chapter 6).

For the managers of university research projects, the main dilemmas are how best to develop expertise in product development. It would appear that universities have become adept at combining basic and applied research as measured by the number of patent applications. Indeed, the business sector has begun to complain about what they consider to be the encroachment of universities into the private sector. But there is evidence that universities are not gaining much from their patent applications, presumably because they have not developed good connections to various businesses that have the correct expertise to develop the patents.

Universities have also been in the forefront of encouraging start-ups via incubator programs for small businesses and have engaged in building industrial research parks near their locations. But the real problem in the construction of linkages with the start-ups or the companies that are located in the industrial park is the creation of the correct kind of diversity in the research teams, capable of handling research on manufacturing and quality control to say nothing about commercialization. Patents are only the first step. One needs to do a considerable amount of research on how the product development can be successfully implemented before the patents can be sold for royalties. Beyond this, just having the industrial research parks next door does not mean that the university will gain that much from their presence without becoming involved in their research. The dilemma for managers of university research projects is not so much that they lack the expertise to do applied research and product development but rather that the faculty is already overwhelmed with too many goals: education, scientific advance, and administration, to say nothing about additional side objectives, such as fund-raising.

One potential solution is for universities to reinvent the highly successful extension model that was used in agriculture but apply it to new areas, especially for stimulating economic development in the low- and medium-tech sectors.[57] This model is discussed in some detail in Chapter 6. Another dilemma for managers of research in universities is that the criteria of promotion have to be changed if faculty is to become heavily engaged in applied research and product development. The potentiality for papers diminishes in product development, in particular where patents are the more likely symbol of progress. One of the reasons to establish a separate division for extension work is to allow for different criteria for promotion to be established.

If universities, because of their size and diverse goals, are likely to have the necessary expertise, small high-tech companies are more likely to find it difficult to have the necessary arenas to create strong connections with other kinds of research organizations, especially if they are located in the middle of the idea innovation network.[58] The dilemma for the entrepreneurs that create them is not only to pull together a diverse team but one that has expertise in enough arenas so that there is overlap with other research organizations in the idea innovation network. Usually, the entrepreneurs who create small high-tech companies worry about the applied research and product development but do not concern themselves with having some expertise in manufacturing research or basic research. It is not so much that they have to solve problems located in these arenas but that they need frequent and intense communication with universities and large high-tech firms that can provide solutions. Given that a new firm is lucky to have only three or four individuals to start with—who are probably working for stock options rather than pay—the entrepreneur has to carefully choose these three or four individuals so that they are diverse in multiple ways. Thus the selected individuals should be diverse not just in their scientific expertise but also in terms of the four arenas that have to be typically represented.

The variability in the number of arenas for commercial success of large business firms in the high-tech sectors is considerable, as we have seen in the comparison in Chapter 1 between the biotech and pharmaceutical sector and the telecommunication sector, and therefore it is difficult to provide more concrete suggestions than the general principle. More examples of differences are found in Chapter 5, where case studies of the automobile and semiconductor industries are discussed in the context of how they did or did not strongly connect their idea innovation networks and supply chains. Also in Chapter 5 is a typology for describing sectors. The key point about large firms is that they have the resources to have the necessary overlap with the arenas that are differentiated in their sector.

Again, usually ignored in books on innovation is the very important and underexploited role of the public research laboratories. These public research laboratories represent a hidden treasure in the United States, and one that distinguishes this country from most other countries with large investments in RDT. For these public research laboratories the managerial dilemmas are different from those of the large industrial firms. Many of the broad-scope research programs in science are not administered in universities but in our public research agencies and mission agencies. While large industrial firms in the high-tech sectors have been moving out of basic research, the large public research laboratories have been moving into more applied kinds of research, as can be seen in Chapter 6. But there is a great deal of variation, and to restore the innovative edge, more needs to be done. Recognizing that many of these laboratories are involved in fundamental science, such as high-energy physics, that for the moment does not appear to have any immediate practical applications, it is also true that these laboratories have developed a number of applied ideas that could be commercialized. Beyond this, the public research laboratories manage a number of what are called user facilities where private companies can conduct expensive experiments. User facilities provide a marvelous opportunity for cooperation between the public sector and the private sector, a point discussed in Chapter 6. The solution for more cooperation requires that the public sector have representation of several arenas besides basic and applied research in their complex teams so that they can connect better with the private firms.

One blockage preventing the kind of overlap that is necessary are the various laws that need to be changed so that cooperation is possible. New legislation should be passed that would allow the public research laboratories to reach out more easily to the private sector that has better skills in manufacturing, quality control research, and commercialization, and in turn for the private sector to have more access to the skills in basic and applied research and in particular the use of the user facilities that the government has built. Therefore, the major managerial dilemma in public research laboratories is how to overcome the opposition of Congress and, more specifically, an ideological view that keeps government and private enterprise separate.

Business firms in the low- and medium-tech sectors face a different competitive situation than those located in the high-tech sectors. Generally, these firms are small or medium in size and have not had a history of industrial research laboratories that one finds in the large high-tech sectors. For them, one can make the case that if the low- and medium-tech sectors can increase the amount of scientific knowledge in their products and services, they will benefit the most in the marketplace. In other words, in contrast to the prescription in open innovation, these firms should be developing their own industrial research laboratories for several reasons. The first and most obvious is that it makes them more capable of absorbing ideas created elsewhere. In particular, this would allow them to develop closer relationships with universities that have the expertise in basic science. But again, the amount of interest that universities will take in their problems is highly variable, which is why I have suggested the importance of developing extension services funded by state and federal governments in Chapter 6.

In summary, the dilemmas, or potential blockages, for managers vary not only in terms of the three dimensions used to construct a typology of diverse research teams but also across different kinds of organizations. Thus, the remedy of broadening the vision of the research team is not as straightforward as it might seem. Only some of the potential blockages have been discussed; there are more.

EVOLUTION AND THE MANAGEMENT OF RESEARCH TEAMS

The various ideas in this chapter represent the synthesis of four social science literatures that do not cross-reference each other.[59] Perhaps the most important insight is found in the work of Jordan (2006) that identified two dimensions for constructing a typology of diverse research teams and indicated some of the dilemmas involved in their management.[60] This chapter has added a third.

The concepts of evolution and failed evolution provide a dynamic quality to the previously cited social science literatures, suggesting that how one constructs a diverse research team must continually change because of the evolution toward more diversity for each of the following reasons:

- To have more scientific breakthroughs and radical technical advances
- To handle more difficult and broad-scope problems
- To ensure overlap between differentiated arenas that are adjacent to the focal research team

Having an evolutionary perspective about diversity represents, I think, an advance in these literatures.

The ideas in this chapter also build upon the concept of open innovation but provide a more detailed framework about where managers should search for new ideas. The six arenas and, of course, their importance shifts from sector to sector, but they provide guidelines as to what information one should be open about. Furthermore, the continual insistence in this book that there is not enough manufacturing and quality research highlights ideas that have been ignored in the thesis of open innovation.

These same ideas provide new insights about the concept of absorptive capacity, which have been mentioned in endnote 56. Simply to measure openness to new ideas by the amount of RDT expenditures is to ignore the barriers of organizational culture that explain the demise of Xerox PARC, Bell Labs, and other industrial research laboratories. More important than the amount of expenditures is the diversity of research arenas that are represented and whether or not research teams, where the information is needed, have technological gatekeepers. Beyond this, the extended discussion of the need for overlap indicates how managers can absorb more from their environment.

Perhaps one of the most important set of insights builds upon the work of Jordan (2006), who observed the problem of dilemmas, or potential blockages, that make it difficult to implement the suggested remedy of increasing the diversity of the research team in order to broaden the vision of the team. Only a certain number of them have been identified and more will be discussed in Chapter 7, where I discuss the importance of feedback about invisible blockages. Again, these are important contributions to the R&D management literature. We need more research on what the dilemmas, or blockages, are at the team and organizational levels and how they can be overcome if we are to restore the innovative edge.

3 STIMULATING THE CROSS-FERTILIZATION OF IDEAS
Step Three in Restoring the Innovative Edge

A MAJOR OBSTACLE that reduces radical innovation is the lack of interdisciplinary communication within research teams. Selecting the correct diverse research team is only half of the managerial task for stimulating radical innovations. The other half, and by far the more critical task, is to have effective cross-fertilization of ideas between the different viewpoints represented in the diverse team so that learning occurs. If diverse research teams are the engines for restoring the innovative edge, then communication that produces cross-fertilization is the energy source that makes these teams work. An important qualification is that the communication is effective, making explicit the hidden assumptions, experimental methods, and intuitive causal reasoning.[1] When this type of communication occurs, radical innovation becomes more possible, assuming that other obstacles have been eliminated. The value or objective of cross-fertilization of ideas is the amount of mutual learning that occurs, as expressed in the following equation:[2]

$$\text{Diversity of Paradigms} + \text{Cross-Fertilization or Learning} = \text{Innovation or New Knowledge}$$

The key component in this equation, of course, is the amount of learning that occurs when individuals with disparate perspectives are team members.

Learning only begins when individuals realize that their present routines are not working. Managers have to appreciate that there are faults or defects in their products and services, and researchers have to recognize that there are errors in their ideas before there can be an effective solution that improves the product or service. Therefore, designing research teams and interorganizational networks so that they learn better becomes a critical step in restoring the innovative edge.

The reasoning in this equation and the importance of cross-fertilization are so central to restoring the innovative edge—at multiple levels—that it is a theme in most of the remaining chapters. However, the focus changes. In this chapter, the focus is on learning within research teams, whereas in the first part of Chapter 4, it is on cross-fertilization between research teams within the same organization. Chapter 5 is about effective communication between organizations in the idea innovation network, while Chapter 6 looks at a special kind of cross-fertilization, the cooperation between the public and the private sector.

Why is effective cross-fertilization of ideas so important? The evolutionary processes described in the previous chapters have put a premium on interdisciplinary communication for the following reasons. First, as we have seen, the diversity of research teams is steadily increasing because of the complexity of the problems that confront the country. Second, the globalization of research and the more rapid development of radical innovations everywhere have made the speed at which radical innovations are developed critical for restoring the innovative edge as well. Effective cross-fertilization of ideas in interdisciplinary teams is the remedy for greater speed. Third, recent policy changes in both the Department of Energy (DOE) and the National Institutes of Health (NIH) are making the amount of learning within broad-scope research teams important. Both organizations have created new research centers to solve difficult and complex problems and are now concerned about how to make them effective.

Nooteboom (1999) was the first to dramatize the difficulty of *scientific understandability* in research teams designed to produce radical innovation. In his original diagram, he defined a dilemma, namely, that as diversity increases, so does the distance between perspectives, which in turn reduces understandability and therefore radical innovation (see Figure 3.1 and its source note).[3] The solution that he proposes for the dilemma created by distance is deceptively easy.[4] Simply accept the optimal midpoint between the decline in understandability and the degree of radicalness, as depicted in Figure 3.1. His solution to this dilemma effectively means that the kind of radical innovation that the crisis in the United States demands is unlikely.

Figure 3.1 represents a modification of Nooteboom's original diagram. It assumes that with a decline in understandability, there is also a decline in communication and, as I shall argue, more critically, a decline in effective

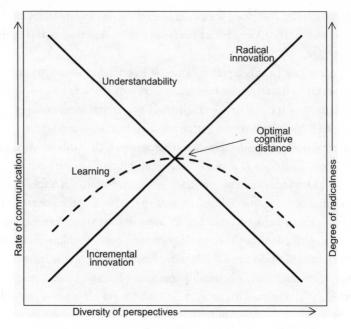

Figure 3.1. A dilemma between diversity and communication
Sources: Nooteboom, 1999: 14; Wuyts et al., 2005; Nooteboom et al., 2007. In the original representation, there are two cross lines, one labeled "communicability" (here labeled "understandability") and the other labeled "novelty." This modification adds the rate of communication on the left y-axis and the degree of radicalness on the right y-axis. "Cognitive distance" is changed to diversity of perspectives.

communication, which is defined below. Novelty is perceived as varying between incremental innovations, or normal science, and radical innovations such as scientific breakthroughs or major technological advances. Consistent with the discussion in Chapter 2, the x-axis, which Nooteboom labeled as *cognitive distance,* here is called the *diversity of perspectives.*

But if research managers decide to reject his solution, what should they do? What are practices for overcoming this natural tendency for effective communication to decline between individuals who have different perspectives? In this chapter, first I set forth a detailed explication of the dilemma and then I suggest three sets of practices: (1) provide information that reduces the distance somewhat between individuals, (2) build emotional bonds that slow the rate of decline in effective communication as diversity increase, and (3) maintain continuous learning that raises the equilibrium curve between effective communication and radical innovation.

THE DILEMMA OF DIVERSITY
VERSUS COMMUNICATION

The easiest way of describing the dilemma is to examine the diagram in Figure 3.1. On the bottom line, or the x-axis, is the measure of the amount of distance within the research team, which is the same as the amount of diversity. In the previous chapter, Table 2.1 presents a list in which each qualitative increase in diversity represents a quantitative increase in distance between perspectives. As I suggested, a simple way of determining this is to examine the overlap in journals or other sources of technical information that are routinely read by the team members.[5]

To increase the radicalness of the innovation—whether scientific break-through or technological advance, whether product or process or both—necessitates expanding the diversity of ways of thinking in the research team or program of research teams, as I indicated in Chapter 2. But as Nooteboom (1999) observes, while combining disparate paradigms creates more radical innovation, as is depicted in Figure 3.1, it also reduces effective communication, the y-axis in Figure 3.1. Precisely because of the diverse ways of thinking, scientists and engineers find it difficult to understand each other's paradigms, especially those aspects based on hidden assumptions and intuitive causal reasoning gained from experience. Hence, the decline in the frequency and intensity of communication between the research team's members. In other words, as the differences in thinking between team members increases, the potentiality for radical innovation also declines, if in fact communication decreases as is hypothesized in this diagram. Hence, the manager's problem of how to increase the cross-fertilization of ideas, the essential reason for creating diverse research teams, given the natural tendency for even scientists to prefer communicating with individuals who share the same paradigm or perspective.

Effective cross-fertilization of ideas has been defined in the opening of this chapter as being able to make explicit hidden assumptions as well as habits in research methods and laboratory practices. But this is not the only component of effective communication. Another is the willingness to take the risk of asking difficult and critical questions, the first step in making the hidden knowledge built upon experience explicit. Another important element frequently requires deciphering the members' nonverbal communication and being able to recognize what is important to them even if they do not verbalize it. Individuals frequently make emotional investments in their ideas or methods because they help define not only their paradigm but frequently their identity as well. These feelings are at a more profound level but precisely one that is not easily verbalized. For this reason, the third section of this chapter focuses on ways of creating emotional bonds between the researchers in teams and even between teams.

Given these barriers to effective communication, how can managers of research teams remedy the situation? The answer is that they must make an extra effort to be sure that cross-fertilization of ideas occurs. What practices can managers use to make an extra effort to create interdisciplinary learning? The diagram in Figure 3.1 suggests at minimum three sets of practices that managers of research projects can resort to. The first and most obvious way is to move the line of understandability to the right so that it intersects the line of radicalness at a higher point. The second section of this chapter describes practices to increase general information: venues to increase training and learning to provide more cross-disciplinary understanding, recruitment of individuals with multiple paradigms, and duo-scientific (different paradigms) leadership of the research team (see panel A in Figure 3.2).

The second and less obvious way of increasing the degree of radicalness is to change the angle on the line of understandability in relationship to the amount of communication. How quickly communication declines given a lack of understanding depends upon the nature of the relationship between members of the research team. If a strong emotional bond exists between the researchers, then communication does not decline as quickly as would otherwise be the case. Changing the angle of the understandability line relative to the differences in perspectives (x-axis) also moves the intersection point toward higher radicalness. Three major methods for building emotional bonds are discussed in the third section of this chapter: eliminating hierarchical or status barriers to communication, creating a knowledge community or a sense of a research "family," and selecting leaders who are emotionally supportive (see panel B of Figure 3.2).

Finally, the third managerial set of practices moves the learning curve higher up so that it too intersects with the degree of radicalness at a higher point. Continuous learning in one sense operates on both the level of understandability and the willingness to continue to communicate even when one does not understand. Clearly, learning in a group situation improves understandability and does so even more than the informational practices previously described. Individuals who learn together also form emotional bonds, which

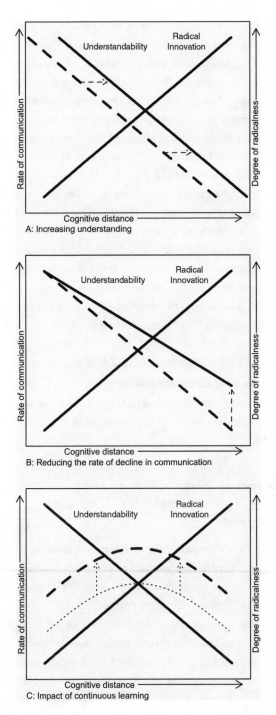

Figure 3.2. Increasing the radicalness of innovation

accounts for the bonds of students after they graduate. Establishing a venue for continuous learning is, however, more complex than it might appear at first glance. Visiting lecturers or staff presentations are common in many scientific and business venues, but their presence does not necessarily mean that learning is occurring. A venue for continuous learning about other disciplines moves the equilibrium upward (see panel C of Figure 3.2).

Just as there are three degrees of complexity in the construction of a research team, we can summarize these three sets of practices for stimulating cross-fertilization as follows: (1) provide information, (2) build emotional bonds, and (3) maintain continuous learning. The evidence for the efficacy of these three sets of practices draws on three different kinds of research studies. The first is an in-depth evaluation of a newly created science and technology (S&T) unit for manufacturing in a large public research laboratory referred to in Chapter 2 as an example of how to build diversity; the second is a detailed case study of biotech firms also cited in Chapter 2;[6] the third is an in-depth study of the Pasteur Institute, a research organization that created a large number of scientific breakthroughs in a short time period.[7]

THE FIRST DEGREE OF CROSS-FERTILIZATION

Practices for Increasing Understandability

What are some positive steps that can be taken to increase understandability? The following three managerial practices are most important for increasing understandability:

- Establish venues for cross-scientific training and orientation
- Recruit researchers with multiple paradigms
- Appoint duo-scientific leadership (i.e., with disparate paradigms) for research teams

Each of these ideas, and in particular the first one, is not necessarily new, but when properly executed it can encourage cross-fertilization by moving the understandability line to the right, as diagrammed in panel A. These methods do not reduce the differences between perspectives because they are not substitutes for the long training and multiple years of experience that researchers accumulate over time, but information bridges can be built that make communication between the disparate modes of thinking easier.

It is common practice for business firms to train potential managers in the business ethos and culture of the firm. Firms take whom they consider to be bright individuals and rotate them among the major branches of the business so that they learn the "language" of each department. In each instance, they spend some time in the department discovering the styles of thinking in sales, production, research or other department within the company. Rotation across the major divisions is a standard operating practice in the military where young officers rotate between the different major departments as they are promoted (e.g., gunnery, engineering, and operations in the U.S. Navy).

The same logic can be applied to the research ethos and culture of the research organization. It may seem strange or unnecessary to send newly minted PhDs through an initial training course, but the objective would be to expose them to the range of fields and methods to be found in the larger organization. This exposure would at least create the basis of familiarity about everyone else's research work, and it is hoped result in a common meta-research language, thus reducing the tendency for a local lab culture to develop. Also, we should not forget that most dissertations are highly specialized pieces of research. Moving back to a broader view of science or technology is not such a bad idea.

The advantages of cross-disciplinary scientific training and orientation courses become especially valuable when a new discipline is being created, as in nanoscience and nanotechnology. A number of disparate specialties are combined and under these circumstances exposing everyone to the techniques and thinking in each of the disparate specialties facilitates their cooperation in diverse research teams. When the Pasteur Institute started the new specialty of biomedicine, it offered a training program. The initial one, which was called the *grand cours*, was not only required of new researchers but it was open to physicians and public health officers from around the world so that there was common language about the new specialty.[8] Some of the more experienced researchers resented this requirement when they first came to the Pasteur Institute but admitted afterward that they had gained a number of insights. The basic course grew to 104 lessons with 23 teachers covering all the various specialties represented in the institute, and it familiarized researchers with the various techniques as new specialties were added in virology, protozoology, entomology, immunology, and so on. This course provided a common background, vocabulary, and most critically, a way of thinking that became recognized as the Pasteurian approach. *When new research organizations are created and they are operating within a new paradigm, the creation of a common language becomes particularly important to facilitate communication.*

Another managerial practice that achieved the same result (i.e., broadening the scientific vocabulary of the researchers) was the practice of moving young researchers from one department or research lab to another, and especially to those headed by Emile Roux (a physician who was superb in developing technical solutions in research) and Elie Metchnikoff (a zoologist interested in immunology). These two research departments were an important part of the young researchers' rotation because they were associated with the most radical innovations (see Chapter 4 for more on these leaders and the consequences for scientific breakthroughs). Obviously, this internal migration simultaneously increased both the integration and the diversity of these departments.

Much the same result occurs when individuals move to different kinds of research problems. The Pasteur Institute encouraged their researchers to examine a range of issues. This intellectual style of studying diverse kinds of research questions had been a characteristic of Louis Pasteur's mind, and it became imprinted in the Pasteur Institute research culture in the early years. Emile Duclaux followed in the same pathway, teaching and researching in chemistry, physics, meteorology, and agronomy, although his most famous work was on the fermentation of cheese.[9] Nor was this atypical, as it also characterized the work of the two other leaders, Metchnikoff and Roux, and was copied by a number of the major researchers, including Améleé Borrel, Constantin Levaditi, Maurice Nicolle, Jules Bordet, and Alessandro Salimbeni (three of whom were foreign, and two of whom made major scientific breakthroughs). Salimbeni attempted to develop vaccines and serums for the plague and yellow fever in his trips overseas, but in his laboratory he studied immunity, serotherapy against cholera, autolysis in microbiology, and pathology of the anatomy during aging.[10] Let me stress that this experience is not the same as being involved in multiple research projects at the same time because these different problems were approached sequentially. This provided an opportunity to learn the new scientific vocabulary associated with the problem.

Such fluidity in the choice of problems and the willingness to tackle them probably strikes many researchers today as impossible in this age of specialization. But fluidity in the kinds of problems is always a question of degree, and probably the degree can be increased much more than is presently found in American research laboratories. The key point is that individuals who move from problem to problem develop a set of behaviors that allows the individuals to become adept at changing, or "a quick study." For example, one approach is to read the relevant literature associated with a specific issue, assuming that it is not too far afield from the current competency of the individual researcher. Within a certain range, individuals can develop a larger understanding and bridge some of the differences in perspective. Such individuals are certainly worth recruiting to be members of a diverse research team.

It might appear that participating in multiple research projects, a common practice in many of the large national laboratories, would produce the same result, but in fact it does not. Indeed, it has almost the opposite result, creating fragmentation and eliminating opportunities for effective communication. Spending a few hours a week working on a project in another department is not the same as spending several years starting with one project and finishing it in the proximity of other researchers who have been trained in a different specialty. In any case, absent a general orientation, the probability increases that while participating with researchers in other specialties, the necessary cross-fertilization does not occur.

Several common practices for stimulating cross-fertilization exist in both research organizations such as the large public laboratories and the industrial research laboratories of firms. Each is a step in the right direction but misses the extra effort that would make it particularly worthwhile. One of these is the house organ, or company newspaper. House organs that report the research findings of different research projects within a national laboratory or a large industrial are also a way that many executives believe can educate people. However, it is important that house organs provide details about the thought processes of the researchers. Including those details would be the extra effort that would make this communication more valuable. A good example of this is P&G's smart learning reports on its extensive Internet system.[11]

Typically in the large public research laboratories, these house organs merely report the findings in one paragraph or two, and *not the process of thinking* or the solutions to technological problems in research, in particular failures that redirected the research, perhaps the most important form of learning. In the case of the Pasteur Institute, Duclaux created a journal, *Annales de l'Institut Pasteur,* that served as the house organ in 1888. This organ provided an opportunity not only for researchers in the institute but others concerned with the same set of issues to report their research and remain abreast of developments in biomedicine.

The most obvious method for managers to increase training and learning is the weekly meeting of their researchers. A common communication barrier is the absence of a venue for the interchange of ideas. Allowing enough time for adequate discussion among the researchers is certainly at the top of the list for eliminating a barrier to communication. In a comparison of biotech firms discussed in the previous chapter, the four highly innovative firms cited the importance of learning more than the four less innovative companies: "We talk as much about how we do work around here as what we work on. The key is to constantly learn from previous experiences and continuously improve our workplace relationships."[12]

But whether discussion among researchers becomes a learning experience depends upon the capacity of the manager to raise questions that force individuals to move to higher levels of abstraction so that the relevance of their detailed techniques and findings can be understood by others working in different areas. Without someone playing this kind of role, these meetings gradually become a form of groupthink. Again, it is necessary to make the extra effort.

This idea of seeking higher levels of abstraction moves us to a larger point about the problems of understandability: Frequently, communication problems occur because the individuals are speaking at a very concrete technical level. But if asked to move to a slightly higher level of abstraction, they are quite capable of doing so, and usually a higher level of abstraction makes communication across cognitive distance more possible. Likewise, if the scientist making a presentation is asked to indicate how his research might be relevant to another person's concerns, then he or she can usually do this. In other words, understandability is increased if the researcher is asked the right set of questions.

An even more effective strategy for increasing understandability than a scientific orientation course is to build a diverse research team with several members that have more than one paradigm. The discussion of the differences between perspectives in Figure 3.1 does not allow for the possibility that some individuals have a broad perspective. Because of their training and life experiences, some individuals have the capacity to think in more than one modality, whether occupational specialty or national culture. Multiple majors produce individuals who can act as bridges between various kinds of diversity, while cross-cultural experiences create researchers more likely to be flexible in their thought patterns and able to read symbolic communication, which is defined in the next section. A number of individuals who changed disciplines from physics to biology, such as Max Delbrück and Francis Crick, gained Nobel Prizes for their contributions. This can be one of the criteria for selection in a research team: multiple majors. In the Pasteur Institute, about one-half of the researchers recruited in the first thirty years had both a medical degree and a degree in biology and therefore were capable of speaking two scientific "languages."[13] Earlier, I mentioned that moving from problem to problem was likely to increase the ability to understand other perspectives and reduce communication barriers; it also implies an individual who has a broad-scope mind. Again, in constructing a diverse research team, it is worth recruiting individuals who have themselves tackled a diverse set of research problems.

An even better indication of a broad-scope mind is someone who has been able to adapt to a new culture other than the one in which he was raised. Given this advantage of learning how to adapt to a new culture, immigrants to the United States, France, and the United Kingdom, as well as elsewhere, frequently win major prizes for their scientific research. For example, foreign scientists (Metchnikoff, Daniel Bovet, Levaditi, and Felix d'Herelle) made four of the eight and one-half scientific breakthroughs at the Pasteur Institute in the first thirty years. In Chapter 2, I argued that one of the greatest signs of diversity is to include foreign nationals on the research team. This is an instance where the problem—cognitive distance created by differences in culture and training—paradoxically provides its own potential solution, namely, individuals with broad-scope minds capable of communicating effectively with a large number of other individuals. Individuals who have effectively learned how to speak two national languages are more likely to be able to facilitate communication between diverse members of the team who speak different scientific "languages." Just as moving from one research problem to another helps individuals become adaptive, so learning how to live in different countries has the same but more powerful effect.

One of the recommendations for increasing the integration of a business firm is the special role of the "middle man," who can be defined as someone whose mode of thinking is halfway between the extremes represented in the research team.[14] Besides emphasizing diverse modes of thinking in the construction of a research team, one can also search for specific individuals whose ways of thinking are interstitial between the extremes represented in the research team. This method for increasing understandability was very much in the minds of the leadership of the Pasteur Institute from 1890 through 1910.

Although public policy debates about science extol the advantages of combining basic research and applied research, which is Pasteur's Quadrant, they have also failed to confront whether there is effective communication between the basic and applied researchers because of the differences in thinking between them.[15] One way of reducing this distance is the addition of a *subspecialty* that is interstitial between basic and applied research, creating some overlap in the journals read in both arenas of research. In the case of biology and medicine, physiology can play this "middle-man" role. The Pasteur Institute recruited Camille Delezenne from Montpellier Medical School to act as a bridge between bacteriology and biochemistry, the basic biological sciences of the time, and the various kinds of applied medical research, for the most part developing vaccines and serums. The implication for research managers is that not all specialties are equally distant from each other. Some specialties can act as bridges between the tacit knowledge of two more distant disciplines.

Perhaps one of the most important but obvious methods for increasing the degree of understandability is to recruit individuals who are willing to work with each other and have the capacity to communicate with each other. The Massachusetts Institute of Technology (MIT) built its new biology department by ensuring that each person that was hired could communicate with each of the other subspecialties.[16] In other words, one increases understandability by ensuring that at least there is a capacity and an interest for cross-fertilization of ideas across a broad scope of intellectual problems. In a relatively short time period, MIT started winning Nobel Prizes in medicine and physiology. Nor is this the only academic department that became successful in this manner. The University of California at San Francisco Medical School decided to create a broad-based molecular biology department with two leaders, one representing basic research and the other representing applied or medical research. The two leaders traveled throughout the country recruiting new individuals, and of course, the critical test was their capacity to think within these quite different domains of thought. Again, this department became famous for the brilliant researchers it produced who won Nobel Prizes for their work.

Another variation on the theme of recruiting researchers who can act as bridges in a diverse research team is selecting all the team members so that they share one quality in common. In the case study described in more detail in the fourth section, the research manager specifically decided to hire a diversity of individuals classified according to their social skills as well as their disciplines (physicists, chemists, and technologists). But besides the diversity, he also searched for individuals who, at least on the basis of their second social trait, would have flexibility. This thinking about the selection of individuals is, as far as I know, quite unusual and indicates a keen awareness of the need to have both diversity and integration within his research unit. The manager worried about competitiveness among researchers and felt that this handicapped the exchanges of ideas. Therefore, he selected researchers who scored second on flexibility, thinking that they would more willingly collaborate. The similarity in flexibility facilitates cross-fertilization. This case illustrates how, in the process of constructing a diverse research unit, one can begin to solve some of the problems associated with the cognitive distance by the characteristics of the individuals being recruited.

The example of the duo-leadership of the basic sciences department in the University of California at San Francisco provides an example of another practice, namely, the advantages of team leadership where the team consists of different ways of thinking. Since National Aeronautics and Space Administration (NASA) and the concept of matrix organization, duo-management is common in research projects. Typically, one of the managers represents the product and customer imperatives, while the other manager represents the technological and supply side imperatives. Leaving aside how effective the matrix style of organization is, I am proposing a variation on this theme, namely, co-project or program scientific leadership *provided that each leader has a different way of thinking*. One of the constant but implicit themes in this book is the need to break many of the paradigms of research and development (R&D) management. One of the simplest ways of increasing understandability upon the part of researchers is to have their team led by two quite different scientific or technological leaders. The dialectic between the two leaders with different specialties becomes one of the major ways of educating researchers in the team to think more broadly and, not inconsequently, to challenge the thinking of research managers. In the process, the researchers begin to appreciate that there are multiple ways of tackling the same problem. The discussions and even intellectual arguments between the two leaders can teach the team members much about how to think. If the two leaders actually work together, then the impact on learning is that much greater.

Again, Roux and Metchnikoff in the Pasteur Institute provided important role models for teamwork by publishing together.[17] The former was a physician interested in applied research, and the latter was a zoologist concerned with basic research. Roux was French and Metchnikoff was Russian. Roux had a critical mind while Metchnikoff was highly speculative. They were original thinkers, which explains why both of them made major scientific breakthroughs and why some of individuals who rotated between their departments also made major discoveries. Many of the researchers who worked with both, either separately or together, marveled at the differences in their thinking patterns and report how this helped shape them in various ways.

In summary, R&D managers can take a series of positive steps to increase understandability, moving the line to the right, as depicted in Figure 3.2, panel A. A common scientific orientation course introduces new researchers to the techniques and thinking of the various specialties within the research organization or the laboratory within an industrial firm, providing the basis for the beginnings of understanding between members of a complex research team. This basis can be supplemented by house organs and weekly conferences. Spending time to recruit broad-scope thinkers and especially individuals who can act as intellectual bridges is well worth the time for later reducing communication problems in a research team. As I have observed, foreign nationals, surprisingly, can provide a considerable amount of expertise in how to stimulate cross-fertilization of ideas. Finally, if each research team is led by two members, each in a different discipline, then as the leaders communicate with each other their team members learn how to do the same. Throughout these three ideas is the single theme of broadening the scientific and technical understanding of the team members so that they can more easily communicate. But, improving understandability is not the only recourse for research managers concerned about increasing cross-fertilization of ideas in diverse teams. Our next topic, creating emotional bonds, is equally, if not more, important.

THE SECOND DEGREE OF CROSS-FERTILIZATION
Methods for Increasing Communication for
Each Level of Understandability

Often overlooked but equally, if not more, important are emotional practices that create bonding between the team members, which encourages effective communication *even with a decline in understandability*.[18] Everyone who has worked in a firm or a research organization can easily understand how we are willing to communicate with those that we consider to be friends and are reluctant to do so with those who are perceived as competitors. Resolving this type of communication problem is where the style of the leader of the research organization or industrial laboratory is so important for impacting on the rapidity with which communication declines given a lack of understanding. Among friends, we are comfortable with admitting what we do not know, whereas with others, we remain more unwilling to do so.

But as I mentioned previously, there are really different kinds of communication. It is not just the passage of information that is critical but also facilitating the exchange of the *hidden assumptions* and the *intuitive causal reasoning* that some authors suggest is essential for stimulating innovation.[19] Making this kind of knowledge more explicit as a form of communication is more likely when there are strong positive feelings or emotional bonds between team members. In effect, when this happens, we change the angle of the understandability line relative to the cognitive distance so that it intersects at a higher level (see panel B, Figure 3.2).

One of the major issues in stimulating cross-fertilization is to encourage individuals to think "outside of the box." The reluctance of individuals to engage in creative thinking is less an issue of understandability than one of taking risks. Scientists and engineers are only willing to take intellectual risks when they feel comfortable, which requires an emotionally supportive atmosphere. The following three practices help to create a comfortable atmosphere:

- Eliminate status barriers to communication and make people feel like equals
- Appoint emotionally supportive leaders
- Create a sense of a knowledge community or research "family," usually with a vision for changing the scientific or technological world

The first practice is again an obvious one, but one honored more in the violation of this principle than in adherence to it. The two other practices, the presence of emotional, supportive leaders and their engendering a sense of belonging to a research family, are also important for creating strong emotional bonds and even a sense of professional friendship between the members of the teams.[20]

The starting point for creating emotional bonds between team members is for the team leader to treat individuals as equals. One source of inequality is bureaucratic rules that confer status on individuals. A commonplace assertion is that bureaucracy kills creativity, but the specific reasons for how this occurs are not usually specified.[21] How does one reduce the heavy weight of bureaucracy? One method is to eliminate many of the rules that interfere with research, a topic that we discussed in Chapter 2. Another and perhaps more critical one is creating a sense of equality by eliminating status differences. Status differences should not be confused with power and influence differences, which I have discussed as one of the managerial dilemmas, that is, of reconciling the goals of the organizations with the objectives of the researcher. Instead, status differences are reflected in pay differentials and in the presence of certain perks. These status differences emerge in large-scale research organizations as ways of rewarding work by increasing salaries and promoting individuals. But, unfortunately, they have a downside to them, namely, increasing the rank differences between people, which in turn make the scientists and engineers less willing to take intellectual risks by expressing wild ideas. Emotional bonds are only possible between individuals who perceive themselves as equals. It is hard to create a sense of a team without a sense of equality between the members and, I might add, between the members and the managers.

This idea was stated by a respondent in the study of high-tech companies, referred to in Chapter 2, who reported: "This organization is different from any other that I have been a member [of]. Discussion is maximized and hierarchy is minimized."[22] These patterns allowed this firm to be a learning organization, explaining how companies are able to reinvent themselves in fields that are rapidly changing, such as biotechnology.[23]

The simplest way of reducing status differences is for team leaders to ask their subordinates for advice about decisions and policies. Research on communication patterns found that when leaders confer downward, the total amount of communication increases—not only upward from subordinate to superordinate but horizontally as well.[24] Although this research focused on communication patterns in the entire organization, it demonstrates how the tone is set by the managers at the top of the organization.

A more drastic remedy for reducing status differences and one that is frequently discussed is the flat organization, as illustrated by GORE-TEX at the beginning of Chapter 2. As effective as this practice is, it can be difficult to maintain given individuals' needs for prestige and signs of superior rank. It means that every researcher has the same size office regardless of the number of publications or international prizes and that salary is more a function of years in the organization than any other criterion. A good example of how effective flatness can be for encouraging widespread communication is that of a major research organization in Switzerland that maintained its size at just fifty major researchers, each with the same size laboratory. They were constructed with an outside passageway so that it was easy for a member of one lab to walk on the outside passageway and talk to a member of another lab. In other words, a sense of equality and the physical arrangements were used to make visible equality as a practice to encourage frequent and intense interactions likely to lead to effective communication. The researchers also all ate together in the same cafeteria. When one of them won a Nobel Prize, this individual demanded a larger office and staff. The research organization refused and he left. Interestingly enough, the message was not understood because when a second researcher won a Nobel Prize, the same problem was repeated and with the same result, the researcher leaving the organization. This kind of organizational commitment to equality is rare but, as one can ascertain, this organization was wise to maintain its policy because it was able to keep making scientific breakthroughs.

The advantages of having a very loose style of organization without hierarchical barriers is demonstrated once again in the case of the Pasteur Institute, which is described in Chapter 4 as an example of an organization capable of changing society. An American chemist who visited the institute during 1925 wrote:

> It is simply a community of scientists drawn together by the incentive of favorable conditions for work rather than pecuniary remuneration ... Some laboratories are classified by the type of research being pursued and others simply under the name of the scientists in charge ... Several persons may be working quite independently upon the same subject. Each one is guided simply by his inclination and aptitude. With few exceptions there are not large groups attacking a given problem under one leader.[25]

In the language of today, it would be described as an extremely flat organization with a considerable amount of autonomy, equality or lack of status differences, as well as absence of bureaucratic rules. The extreme independence of the Pasteur Institute is probably neither possible nor necessarily desirable.

A research manager might ask, How should one handle this problem in a large research organization, university, or industrial laboratory? The large size already means that there are major status differences, to say nothing about bureaucratic rules. Yet, one potential solution would be to create a largely autonomous subunit in which the researchers that are recruited are all approximately the same rank. This characteristic is called "skunk works" in the management literature on innovation and is so described in the prize-winning book *The Soul of the New Machine*, which deals with the development of a new computer.[26] However, skunk works do not have the official sanction of top management, and this device is unnecessary when subunits can be created. New subunits can be effective for radical technological advances provided they are granted considerable autonomy in decision making and freedom from most of the bureaucratic rules that have accumulated in the larger organization.

Another remedy for eliminating status differences is to avoid hiring stars. The study of eight high-biotech companies cited above observed that the recruitment of top stars hindered communication.[27] In the study of the Pasteur Institute, the recruitment of the "best and brightest" from the *grandes écoles* did not result in any prizes for significant research. Instead, it was the individuals recruited from the less prestigious schools as well as the foreigners that accounted for the great breakthroughs that were responsible for the Golden Age in this research organization. The reason is that frequently stars are not good team members. Beyond this, the typical way in which individuals advance in graduate schools, especially in the European case, is on the basis of their intelligence and *not* their creativity. The issue is to recruit creative individuals who think "outside the box" and therefore might not shine necessarily in academic studies, where thinking "inside the box" has been encouraged. Because creativity is not usually measured, especially in the elite schools, recruiting beyond them is usually a good idea.

Still a third remedy for eliminating status differences is to reward young people with good ideas. As Roux, when he was head of the Pasteur Institute, put it: "In our house, we measure the importance of a man from the services he affords and the only title for promotion is scientific production. We welcome whoever brings forth interesting ideas."[28] This encourages individuals to take risks in their ideas.

The normal image of a leader is someone who has a vision and strategies for realizing that vision, which emphasizes the rational side of leadership. Less common is the image of the leader that creates and supports emotional bonds between people. Yet, it is extremely important, especially for encouraging frank conversation between people. Unfortunately, this topic is seldom discussed in studies of radical innovation.

An important managerial practice for building bonds between members of a research team or between teams within a research unit is for the leaders to be emotionally supportive. What does this mean? A good example is the leader of the new research unit discussed at length in the next section, which illustrates three ways of being supportive: mentoring, praising ideas, and resolving interpersonal conflicts. Patten (a fictitious name) demonstrated mentoring not only in his many discussions with the new members of the team about the choices of their research projects but his encouraging researchers to take courses and even work for additional degrees. This policy of furthering the careers of individuals led to considerable commitment upon the part of researchers and created a family atmosphere. The symbolic importance of his support occurred when Patten made no comment when the researchers missed work because of their classes.

Many of the other researchers commented that Patten shared research progress with everyone in the larger organization so that there was public recognition of the entire unit's accomplishments. In this way, he praised his research workers, providing them with a sense of respect, which is critical to researchers.

Again, Patten helped build emotional bonds in another interesting way. One practice of his was encouraging each individual who had some conflict with either another staff member within the department or one outside the department, to come and discuss it with him, including what might be potential solutions. In other words, rather than the manager providing the solution, he tried to guide his staff to recognize what they themselves could do. Doing so also created a sense of equality, especially as a number of the conflicts occurred with one of the older individuals who had been recruited into the S&T manufacturing unit.

During the interviews, the researchers commented on Patten's ability to understand them emotionally, reflecting his ability to read nonverbal communication. One respondent said that Patton showed emotion himself when discussing people and their personalities. Another researcher stated: "[Lex Patten is] good at figuring out where people are coming from." This empathetic quality is a critical aspect of the kind of supportive leadership needed to build emotional bonds.

The building of emotional bonds between researchers is also more likely when leaders concern themselves with the families of the researchers. Roux, the director of the Pasteur Institute, would make a point of talking to each of the children of the researchers who played on the grounds of the institute and received widespread affection from them.[29] He also advised Eugene Wolman to travel to Chile for several years to earn enough money so that when he returned to France, he could afford to buy a house. (At this time, during the 1920s and 1930s, the salaries at the institute had fallen behind as a consequence of the rapid inflation produced in the early 1920s.) Wolman took his advice, but when he returned he was sick. Roux, who was over 80 at the time and ill himself, would get on the metro in Paris and travel across the city to visit him in the hospital.

One of the key ways in which research managers can create emotional and intellectual leadership is reading nonverbal communication in group meetings. How does one read nonverbal communication? This is easier to do than one might think. Primarily, it means paying as much, if not more, attention to *how* things are said as indicated in the tone, gesture, body stance, and timing of the communication than sometimes to what is said. And sometimes the silences are more telling than the moments of conversation. In general, one wants to observe and be sensitive to the level of excitement and how genuine it seems when individuals express ideas.

Individuals vary in what their silences mean and how they express disapproval. In particular, for managers it means learning how the various members of their team who want to criticize express their disagreements without verbalizing them.[30] Let us not forget the differences are even greater by culture, one reason why combining different national cultures in a diverse team can create a considerable barrier to communication; we do not necessarily understand the different meanings of silence. A large literature exists on cultural differences. Here I will provide only a few examples. Upper-middle-class English men lean forward in their seats; how far is an indicator of how strong the disagreement is. The Japanese, as is typical of many cultures that avoid conflict, will look at the ground and say nothing. On the other hand, Americans take silence as agreement. For this reason, it is particularly important to observe silences and attempt to differentiate when it represents disagreement that individuals are unwilling to express.

In the Introduction, I stressed individuals should be encouraged to take risks when they explore ideas. When there is a strong emotional bond, it is both easier to suggest risky ideas *and* to criticize them. One way of creating this emotional bond is to create a sense of belonging to a research family. Then the researchers cooperate in transforming the ideas of each other so that they become better. A research family or knowledge community can be defined as "a group of individuals who have learned how to communicate honestly with each other, whose relationships go deeper than their masks of composure, and who developed some significant commitment to make others' conditions their own."[31]

How does one make a researcher feel like a member of a knowledge community or research family? One organizational policy that can help create the sense of belonging is investing in the professional development of their researchers. IBM created very strong loyalties to their firm by providing, without asking many questions, considerable choice in further education.[32] Investing in researchers' educations is treating them not only as members of a particular knowledge community, but treating them symbolically as members of a research family attached to an organization. The biggest sacrifices that most families make for their children are to invest in their education. When companies do this as well, considerable loyalty is developed. Previously, I observed how important Patten's mentoring of the new researchers was for encouraging creativity.

Other ways that exist for creating the sense of a knowledge community include the focusing on a shared problem considered important. A shared vision can strongly unite the researchers. Monod created one around the emerging new paradigm of molecular biology by ensuring that anyone who was doing interesting research in this area would pass through Paris and give a talk on their latest research. Cohen, one of the researchers that attended these meetings, described them as his postgraduate education.[33] Furthermore, the conversation was facilitated by Monod's sharp questions to encourage critical thought as well as look for solutions to meet the criticisms.

After the talk, all would wend their way to a French restaurant for a good meal and even better conversation. In this atmosphere, questions and "half-baked" ideas could be easily tossed around. It is necessary to emphasize the role of meals in creating a sense of family and opening avenues of honest communication. As is well known from the history of the discovery of RNA, Monod, Jacob, and Lwoff along with their researchers would have daily lively lunches in which no topic was out of bounds.[34] Yet, another example is the kinds of conversations that occurred at lunch at the California Institute of Technology (Caltech). Delbrück, who was famous for how severe a critic he could be of the ideas of his research team, at the same time generated a sense of a research family. He instituted the practice of having his research team at Caltech travel to the desert for relaxation during the week along with his wife and various significant others of the research team. By creating this sense of one's lab workers belonging to one's research family, criticisms are more easily accepted and made. One feels secure within the context of the micro knowledge community.

Besides investing in professional development and involving the research teams in bonding activities such as meals and picnics in the desert, the leaders of the research organization can express personal interest in people as though they were members of the same family. Albert Calmette described the Pasteur Institute as a temple dedicated to science, in which Pasteur surrounded himself with his research family in an atmosphere of work, of total commitment, and of *tenderness*.[35] This direct translation emphasizes a number of incongruities, at least for Anglo-Saxon researchers: First, it is a family working together; second, there is tenderness; and third, the concept of an emotional bond and support is central for the integration of diverse ideas. At various points in his letters home, Calmette describes the encouragement he received, the warm welcome plus the way in which the family atmosphere manifested itself.[36]

In short, reducing the rapidity of the decline in communication rates by creating emotional bonds via the elimination of status differences, supportive leaders, and establishing a sense of a research family is probably more important than moving the line of understandability to the right. The major advantage of emotional bonds over cognitive understanding is that they encourage individuals to take risks and to freely criticize other individuals' ideas. Through these frank exchanges, individuals are more able to make explicit the hidden assumptions and habitual methods involved in what they know, which creates the learning that leads to radical innovations.

THE THIRD DEGREE OF CROSS-FERTILIZATION

Creating a Venue for Continuous Learning

Another and perhaps the most important managerial practice for overcoming barriers to communication caused by differences in paradigms is continued learning.[37] As is apparent in Figure 3.1 and panel C of Figure 3.2, continuous learning moves the equilibrium curve higher precisely because it impacts on both understandability and reduces the rate of decline in communication given differences in perspectives. The idea of continued learning recognizes some of the properties of how we learn. Generally, learning occurs through bits and pieces—the "Aha!" experience is relatively rare—and it is only slowly over time that one gains the insights that makes for a more complete understanding. This third set of managerial practices for cross-fertilization recognizes this temporal dimension and is therefore different from the first set of practices, which involve simply providing more information. In the first set of practices for increasing understandability, the creation of a scientific orientation course, represents a one-shot injection of knowledge. Creating a venue for continuous learning assures that this process endures and is much more likely to lead to a deeper understanding of different disciplines and paradigms as information about other perspectives accumulates across time.

The importance of continuous learning is illustrated in the following description of how the Wright brothers went from building bicycles to creating the first airplane. Although not an example of a diverse team—the two brothers lived, slept, and thought together their entire lives—it is an illustration of how they continued to learn.[38] They first spent four years studying the patterns of bird flight and everything that was available on the flight of gliders. Then in 1900, they made multiple trips to Kitty Hawk and after each flight modified the wings of their glider, learning how best to take advantage of the wind patterns. In 1901, in the next season at Kitty Hawk, they realized that the wings were not providing enough lift to carry the motor that they would eventually mount on the glider. So in Dayton, they built a wind tunnel six feet long and with a powerful fan tested two hundred wing designs. In 1902, on their third trip to Kitty Hawk, they discovered still another problem that had to be solved, the yawing of the glider. They experimented with various solutions, first trying a vertical tail and then realizing that it had to be moved to the left or right with a control from the pilot. Powered flight finally occurred in 1903 after several months of fixing small problems. This story illustrates how multiple problems have to be solved before a radical innovation takes place.

Another reason why continuous learning is important is because as science and technology advances, what we know becomes less and less relevant.[39] In a fast-paced, rapidly growing body of research findings, the frontier recedes quickly. This problem is different from that discussed with the creation of new designs that make previous competences obsolete. Instead, it is how quickly what we know becomes outdated. I have already reported the example of the disk drive. Each technological leader was quickly replaced with another as the disks became smaller and smaller. The reason for this decline was too much attention to the customers and thus productivity and not enough to the rate of technological change and the idea of growth through new products, which was the strategy of Hewlett-Packard, reported in Chapter 1.[40]

To provide managers with a set of practices to create a continuous learning experience for their research, I draw upon an interesting and relatively rare case study of the creation of a new department, in which individuals are learning from each other and from other researchers in other departments that were located in a large public research laboratory. One of the themes in Chapter 1 was the importance of emphasizing manufacturing research. The relationship of basic scientific research to manufacture is well known in semiconductors (see Chapter 5) but has been underappreciated in some of the other high-tech sectors where it is equally important (see some examples in Chapter 6). Basic scientific and applied research on manufacturing processes has been a relatively unexplored area in the research management literature, and yet for restoring the innovative edge, it is essential. In Chapter 2, I used the creation of a new unit to focus on the science and technology of manufacturing to illustrate all the ways in which diversity could be defined. In this chapter, our focus is on the various practices that Lex Patten, the fictitious research manager, employed to create a situation of continuous learning.

One of the key decisions that Patten made for building a knowledge community and encouraging continuous learning was the involvement of his new researchers in projects in other departments, thus building opportunities for cross-fertilization across departments. This process of connecting his new researchers with others in other departments was facilitated by his extensive set of contacts, given the number of years that he had been working for this large national research laboratory. Because these other departments had their own subcultures and modes of thinking, Patten tried to reduce the distance by inviting those involved in research projects that are located outside the department to make presentations to the entire staff of the S&T manufacturing unit. These presentations provided a venue for effective communication.

Lex Patten created multiple complex research teams with a range of networks built around joint research projects. He avoided having all team members within the same unit. Instead, he built a knowledge community that transcended the unit to involve individuals in other departments, thereby creating a knowledge network. He recognized that having individuals outside the unit would result in more learning than simply hiring double or triple the number of individuals and would provide the opportunity for continuous learning. It would also reduce the cost of hiring two to three times the number of individuals. But given this design, the issue was ensuring involvement and commitment to the joint projects in this knowledge network. He arranged to have about 30 to 40 percent of their salary paid over a three-year period. In this way, their involvement in the knowledge community was guaranteed for a long enough time period to make the investment worthwhile for both the individuals involved and the S&T manufacturing unit. Over three years of presentations in the meetings by individuals who continued to work in partnership with members of the S&T unit led to the gradual exposure of their hidden knowledge and intuitive reasoning.

Most of the researchers within the department found that these meetings were intellectually exciting. The meetings became a critical managerial practice for encouraging cross-fertilization between the various projects within the unit, as the common knowledge pool was raised during the presentations and the question-and-answer periods that ensured. In other words, the physicists were learning not only about chemistry projects within their unit but chemistry projects located in other departments, and vice versa for the chemists. One respondent who had worked in this national laboratory for a number of years stated that there was much more cross-fertilization in this new unit than in the previous two departments that he had worked in.

Why is this design brilliant? A large literature in organizational sociology and a much smaller series of studies of research organizations indicate that as size grows, there is a tendency for internal differentiation to occur. In this instance, if more physicists or chemists or engineers had been hired within the unit, there would have been a natural tendency for those of the same discipline to form subunits, hurting cross-fertilization across these disciplines. Those principles of differentiation discussed in Chapter 1 operate within units as well, in this case with negative consequences for interdisciplinary communication. Furthermore, the knowledge of all the researchers is increased much more by having joint projects across unit boundaries because this practice provides more access to the diversity of projects and thus the pool of knowledge in the other units that are now attached to this unit. This practice draws upon the same logic provided in Chapters 4 and 5 about the advantages of external alliances as a source for expertise. But for these advantages to be realized, it is necessary to provide a venue of communication in which continuous learning occurs and both internal and external communication are maximized. Thus another key element was the presentations each month not only of research involved in joint projects but also other research that the individuals were involved in.

Nor did the creation of a knowledge community and venues for communication stop with these monthly meetings and presentations. Patten worked to develop close relationships with the design department as well. According to one respondent, he did so by avoiding competition and broke down the barriers between the S&T manufacturing unit and the product design department by asking the right questions and talking to them about their work. Because of this effort, the S&T manufacturing unit had some influence in the selection of a basic researcher who was hired into the product design department. In turn, this selection led to the development of the colocation of some research equipment that benefited both departments. Colocation is a particularly useful technique for continued learning. Patten was described as having "pulled strings so that a member of the group would share an office with a member of [S&T manufacturing unit]." This move created a considerable amount of consulting about the next generation of product design.

One of the fruits of Patten's efforts was that two scientists, one in the S&T manufacturing unit and one in the design department, began to organize semiannual conferences for others interested in the same set of problems. These conferences obviously broadened considerably the boundaries of the knowledge community beyond this one unit and national research laboratory.

But, to keep any knowledge community fresh with new ideas requires that it move beyond the walls of a single unit. This case study illustrates a variety of ways in which the knowledge community was expanded via collaborative projects with other units, with semiannual meetings with others in other organizations. and by close ties with the product design department. But most important was the venue in which presentations by others in other departments became the topic of conversation and the basis of continued learning in which the scope of each individual's knowledge was expanded and thus distance was reduced.

One of the arguments that I have been making is that continuous learning broadens the scope of individual researchers so that the equilibrium point moves higher. Some evidence for this comes from one of the interviews where the individual stated he was finally beginning to understand the whole scope of the problem that needed to be solved because he was now involved and interacting with researchers focused on other aspects of the problem. Clearly, having a "big picture" or vision, which is created via the continuous learning, facilitates communication.

The three sets of practices described in these three sections solve the dilemma between diversity and effective communication. If research managers apply them, they can increase their rate of radical innovation. The first set attacks the communication obstacle by moving the line of understandability with better information about other perspectives or individuals who can act as mediators; the second set focuses on creating emotional bonds so that the decline in communication, given differences, is less great; the third set focuses on moving the equilibrium between understandability and radical innovation higher. Together the practices discussed in each section provide clear guidelines for managers on how to generate cross-fertilization and manage the processes of innovation at the research team level.

EVOLUTION, EMOTIONS, AND THE ORGANIZATIONAL KNOWLEDGE PARADIGM

In Chapter 2's conclusion, I observed that four social science literatures can be integrated and combined with the ideas of evolution and failed evolution. Therefore, the three sets of managerial practices—providing more information, creating an emotional bond, and establishing venues for continuous learning—make a contribution to these same four literatures and provide a remedy for the lack of interdisciplinary communication. The three sets of practices also allow us to revisit the ideas of increasing the absorptive capacity of the research team and making it more open to ideas outside the organization.

An important implication of the increased emphasis on the speed in the creation of radical innovations and the shorter product life cycles implies that R&D managers should increasingly emphasize the second and third set of managerial practices. Strong emotional bonds and venues for continuous learning are likely to facilitate speed.

In this chapter, we have added to these social science literatures an important cognitive perspective for economics as well as sociology—that is, emotions—and a new one for management—that is, the organizational knowledge paradigm or the learning organization.[41] In the latter literature, a key problem is how to create integration between the disparate perspectives.[42] As research projects grow larger in scope and the problems become difficult, then the organization has to become a learning organization, as has been indicated in the project development research.[43] It is no accident that the research topics within the social sciences have also evolved as part of the evolutionary process described in Chapter 1.[44] These developments represent attempts to deal with the new issues posed by these evolutionary processes.

For the literature on the sociology of emotions, these practices add the obvious insight about the importance of emotionally supportive leaders and the less obvious one about the importance of creating research families within the work context. An important idea is the reading of nonverbal communication as way of understanding hidden feelings that are not being expressed so that they can be dealt with.[45] I would argue that this idea is the key to supportive leadership and integrating diverse perspectives.

For the new specialty of the learning organization or the knowledge paradigm, a major issue is the importance of tacit knowledge, the hidden knowledge gained from experience. The whole chapter has focused on how to tap into these wells of knowledge gained from experience. To my knowledge, the learning organization perspective does not have as yet much about managerial practices for encouraging learning within research teams, an obvious topic of this chapter. But perhaps the most interesting set of insights being added to this perspective is the importance of emotions for creating more effective learning.

4 INTEGRATING THE ORGANIZATION AND CHANGING ITS CONTEXT

Steps Four and Five in Restoring the Innovative Edge

THE TWO PREVIOUS chapters focused on various types of diverse research teams and multiple practices for generating cross-fertilization of ideas. These research teams do not exist independently but are housed in organizations ranging from universities to national research laboratories and from small high-tech firms to large companies that engage in industrial research. A major obstacle that reduces the rate of radical innovation is the tendency for "silos" or "stovepipes" to develop around different departments within a large organization. This development interferes not only with the integration of departments within an organization but also with establishing relationships with external organizations.

In turn, these organizations also exist in a larger societal context that impacts on their ability to create radical innovations. At the level of the society, there may be certain obstacles that prevent an organization from making scientific breakthroughs or radical technical advances.

This chapter focuses on two obstacles, the lack of interdisciplinary communication between departments or divisions within organizations and the tendency for managers to be reactive toward changes in their context rather than attempting to change society and even solve global problems. For both obstacles, a key remedy is transformational team leadership. Unlike the previous two chapters that examined obstacles at the team level, these obstacles are located at the organizational level.

The first obstacle, the lack of cooperation between departments or divisions as well as between departments and the external society, has received a name recently in the policy literature and that is the tendency for "stovepipes" or "silos" to develop. A recent report from the Brookings Institution stated that the Department of Energy's (DOE) national laboratories were "too siloed" to move quickly to commercialize their research.[1] Another report from Harvard's Kennedy School argued much the same, stating that DOE need to forge better links between basic, applied, and deployment. The criticisms are not completely fair because some of DOE's national laboratories have been developing products jointly with industry (see Chapter 6). Furthermore, the problem is a common one in large organizations and is still another reason for the downfall of General Motors as well as for the trouble some universities have in developing interdisciplinary programs (see Chapter 5). Because departments compete for resources, they are generally unwilling to cooperate to solve larger national and global problems.

To remedy the first obstacle, three organizational practices—(1) complex strategy, (2) team leadership, and (3) different sources of revenue—are identifiable that create cross-fertilization between different departments and divisions within the organization. An extended example in the Pasteur Institute of the benefits of these practices is reported in the first section of this chapter.

The second obstacle—the tendency for managers to react to their environment rather than be proactive in solving national and global problems—has been demonstrated in example after example in the Introduction. As we have seen, the emphasis on productivity and stock prices allows complacency in a changing world that needs more innovation. What is needed are proactive leaders and, even better, a transformational leadership that attempts to change society. The signs of transformational leadership is an articulated vision of what can be. The case study in the second section provides an example of this kind of leadership in the Pasteur Institute. But the Pasteur Institute also provides lessons about how to overcome a number of obstacles within society as it attempted to change French public health and find new treatments globally. It is this visionary aspect that is most informative.

Why choose this specific historical case? From 1890 to 1920, the Pasteur Institute had eight and one-half radical innovations, literally qualifying as a burst of scientific breakthroughs. This number was more than in any other research organizations during a thirty-year time period throughout the one hundred years that were studied in what amounts to a quasi-census of radical innovations in biomedicine.[2] The institute created the transdisciplinary area of research we now call biomedicine, which represents the combination of basic and applied research, currently a policy issue given the interest in creating more new interdisciplinary specialties such as nanotechnology.

Finally, why are Steps Four and Five combined in the same chapter? It is more than just the use of the same case study of the Pasteur Institute. One major reason is the central role of transformational team leadership in overcoming both obstacles. In addition, the two other organizational practices—(1) a complex charter and (2) multiple sources of revenue—are equally relevant. The former provides a mandate for societal change—a strategic objective—and the latter creates the surplus funds to work globally.

MANAGERIAL PRACTICES FOR INTEGRATING RESEARCH DEPARTMENTS

Complex Charter, Transformational Team Leadership, and Diverse Funding

As suggested in Chapter 3, the problem of interdisciplinary learning via cross-fertilization of ideas is the theme throughout multiple chapters. That chapter discussed the problem of learning between members of the research teams. In this chapter, the focus is on learning between different departments and divisions within the organization. At this level, the problem of cross-fertilization is more difficult. And as is indicated in this section, the three organizational practices that encourage learning did not work in all instances and the failures are instructive.[3] In Step Four for restoring the innovative edge, however, the issue is not just learning but, perhaps more critically, cooperating to achieve a larger goal of changing society, which is the topic of the next section.

Three organizational practices—(1) complex charter, (2) transformational team leadership, and (3) diverse funding—encouraged cross-fertilization and cooperation between several departments but not all. We can verify the importance of these three practices by comparing the Pasteur Institute in the period of the burst of radical innovation with the next thirty-year period when most of these practices have fallen in disuse, with a corresponding fall in the rate of radical innovations.

Most research organizations begin with a specific mission statement, or raison d'être, that represents their intellectual agenda, or what some might call a *charter*. The charter establishes a template that influences the trajectory of the organization as it evolves across time. The vision of the founder is expressed in this charter, and in this regard the Pasteur's vision was truly exceptional in comparison to the equivalents at other institutes founded both before and after it. What makes this charter exceptional and worthy of emulation were the following three characteristics:

- It focused on a number of national and international problems.
- It combined basic science with applied science, creating the new specialty, biomedicine.
- It added the four other arenas of the idea innovation network.

Why is the charter so important for overcoming stovepipes? It indicates a larger vision, indeed even a heroic one, that moves considerably beyond the mundane concerns of petty departmental concerns. In Chapter 3, I stressed the advantages of creating a knowledge community. This is the same idea but on a much broader scale. Furthermore, the focus on national and global problems gives the charter an emotional component that leads individuals to work day and night and not to care about salaries, size of lab, or other perks.

What were the components of the charter at the Pasteur Institute? First, the charter laid out the very broad objective of creating a whole vast new area of inquiry: biomedicine. Rather than beginning with a narrow focus on bacteriology, as did many of the other new research institutes in biology founded at about the same time, Pasteur's broad-scope strategic vision combined the basic science of biology with the applied science of medicine. His model of good science was a general theory, Newtonian physics. At the time, biology was largely dominated by descriptions of the differences between plants or between animals or other biological species. The only general theory available was, of course, Darwin's theory of the evolution of species, one not relevant to the problem of biomedicine at the time. Pasteur's vision thus coupled two areas, basic research and applied research, that were generally not combined in France or elsewhere, especially in the 1880s. Indeed, the term *biomedical research* did not become common until after the Second World War.

Furthermore, Pasteur's larger objective was the methodology of experimental medicine, which was concerned with generating new knowledge *for public health*. Pasteur was *not*—unlike many of his French contemporaries and successors—taking an ideological approach to his *theoretical* view on bacteriology, but instead arguing for experimental medicine, essentially a *methodological* strategy, an extraordinarily broad and open agenda. Unlike the German school of bacteriology, which focused on the *causes* of infectious diseases, the French school or, more particularly, Pasteur and his disciples were more concerned with finding the *treatment* for these diseases.[4] This might appear to be the same objective, but there is a subtle difference. The German school believed in universal causes, whereas Pasteur's model recognized that the human body influenced whether the cure was efficacious. In other words, in the latter's view there was not a universal treatment, except in a few cases. This same perspective led the French researchers to be concerned about how to improve the treatment, which is discussed later in this chapter. In other words, a century before it become common, the idea of customization of treatments is subtly introduced. As I noted in the Introduction, the advantages of combining basic and applied research have been labeled Pasteur's Quadrant in the policy literature, another reason why this particular case history is so important today.[5]

Second, the charter contained in its strategic vision the goal of changing the nature of public health through the education of physicians in the techniques of medical research.[6] Why was this so important? At the time, French physicians adhered to the wrong model of how to treat illness; they relied on the vitalist tradition.[7] The emphasis of the new training programs was on the teaching of techniques and methods in medical research to doctors, so that doctors could understand better the causes of diseases and how to think about them. Perhaps the best indication of the importance of this objective was Roux's original plan for the new building built with funds from the first public subscription, in which one-half of the space was allocated to teaching.

As if these two strategic objectives, creation of a new interdisciplinary specialty and changing public health in France, were not enough, the institute embarked on a third, the mass production of vaccines and serums in order to create financial independence from the state.[8] Again, in the context of France, this was revolutionary and critical for maintaining independence in the nature of research. Pasteur started this strategy by giving his royalties to the institute, a precedent that was followed by Roux.[9] In particular, Roux's achievements in developing the serum for diphtheria provided a steady stream of income because the state financed the inoculations of all the school children.

Separate laboratories were established for producing and improving the quality of each vaccine or serum, creating another first: the built-in research agenda for constantly improving the quality of the vaccines and serums, a major issue in the beginning. As I noted previously, part of the reason for this concern emerged out of the desire to develop treatments while recognizing that there was no perfect one. Hence the need to continually improve the treatment in so far as this could be accomplished. This research was conducted in diverse teams that connected basic science and applied researchers with the researchers in the production laboratories. The former created new toxins that were then sent to the production laboratories, which in turn produced the antitoxins and serums. *Thus, the Pasteur Institute became the first research organization to also become an industrial research laboratory.* But this source of income supplemented other sources, as is indicated later, providing financial independence to the institute.

The combination of basic and applied research with concerns about quality and production meant that the complex charter or multiple strategic focus of the institute considered all parts of the idea innovation network and was, in effect, similar to the kinds of closed research laboratories just being founded at General Electric and AT&T in the United States at the same historical moment, having been preceded by Siemens in Germany. Even the way in which the vaccines and serums were commercialized was carefully considered. Large-scale production of diphtheria vaccines were ordered, purchased, and distributed by the state. In contrast, rabies shots were handled in separate clinics in different parts of France and of course elsewhere because of the need for rapid inoculation. Animal products such as a virus for killing rodents were sold for profit, while human vaccines and serums were sold at cost.

In summary, the strategic vision created a new specialty, biomedicine, confronted the wrong treatment model of the physicians, attempted to train them so that they would focus on the causes of diseases, focused on improving the quality and production methods for vaccines and serums, and helped establish financial independence. Thus, it combined all the major arenas of research found in the idea innovation network. It had the national goal of changing French public health and the global perspective of finding treatments for various diseases. This broad strategic vision provided a way of uniting the different perspectives in the disparate divisions of the Pasteur Institute, allowing them to easily work together—if they so desired—united by the same vision. It created a sense of a research family, and they referred to themselves as "the Pasteurians."

A second unifying element was the transformational team. Why is a team, especially a transformational team, better than a single director or a CEO? Generally, leaders tend to favor their own discipline, as did the accountants in General Motors, and the same happened in the Pasteur Institute. But as we shall see, their biases created a diverse intellectual body of researchers and this diversity as well as the diversity of the leaders generated an atmosphere of cooperation, especially as the leaders did research together. The team integrated in various ways or tried to—for it is important to recognize that in substantial ways they failed—the different research teams. As the leadership team attempted to handle national and global problems, their efforts again provided substance to the vision, uniting the disparate divisions of the Pasteur Institute. This vision helps overcome stovepipes.

Another advantage of the transformational team over the transformational leader is that it can offer distinctive visions about how to change society and what obstacles need to be confronted. This was the case of the team in the Pasteur Institute. Precisely because their visions about both internal and external integration differed, this diversity in vision greatly enhanced the capacity of the institute to create a number of scientific breakthroughs, eliminate different obstacles, or resolve blockages in their research environmental context, thus altering the world of science dramatically—at least up until the First World War. In this section, the emphasis is on how they integrated various departments, while in the next section, we will shift to their visions for changing the research environmental context.

Who were members of the leadership team? Surprisingly, not Pasteur, who very quickly fell ill two years after the founding of his institute and died in 1895. Without denying the importance of Pasteur and his influence in the creation of some of the more critical aspects of the charter, it was the team leadership of Duclaux, Metchnikoff, and Roux that propelled the Pasteur Institute during the 1890s and 1900s.[10] Roux already was implicitly acting as director from the institute's inception in 1888, even though Duclaux had been chosen as Pasteur's successor.[11] One of the aspects that made team leadership possible was that all three of them were intimate friends and had respect for each other's scientific and medical research.

In Chapter 3, I reported that Metchnikoff was a zoologist with a speculative mind while Roux was a physician who was superb at the technology of research. To complete the description of the differences between the members of the leadership team, Duclaux was a physical chemist and engaged in applied research in areas other than in medicine, mainly concerning agriculture and industry, but like Roux he tended to have a critical mind, doubtful of hypotheses. These differences in disciplinary backgrounds, research experiences, and cognitive styles led to different visions about both internal and external integration and influenced the recruitment patterns that created the diversity for the construction of complex research teams.

Again, let me emphasize that perhaps the most decisive way in which senior managers can increase the radicalness of their innovations is via recruitment. So, the advantage of having team leadership becomes apparent when examining the different recruitment patterns of the three leaders. Duclaux not only drafted Roux but even paid for his medical studies.[12] Obviously, this is a very effective way of creating the sense of a research family. When G. Bertrand made his discovery of enzymes, Duclaux immediately hired him and quickly promoted him to be a *chief of service* (or department head, a distinctive title at the time) at the institute.[13] Duclaux was also very active in recruiting August Fernbach, Pierre Mazé, Alexandre Etard, August Trillat, Henri Mouton, André Staub, and his own son.[14]

As for Roux, he engaged Calmette, Yersin, and Louis Martin in medical research in the very beginning of their career. Then, when he learned that Laveran was retiring early from his military career, he offered him a laboratory at the Pasteur Institute in 1897, which led to the inclusion of tropical medicine in the study of microbiology. Another critical choice of Roux's was Fourneau, who came with a whole new strategy for chemical therapy in 1911 that led to the Nobel Prize–winning work of Bovet.[15]

Finally, Metchnikoff saw Borrel's thesis (medicine) in Montpellier and enrolled him in his laboratory. He also asked Albert Berthelot (a physician and a chemist) to work in his laboratory. In addition, he attracted a number of foreigners, including Russians (Alexandre Besredka, Michele Weinberg, Wollman), Romanians (Jean Cantacazène, Levaditi), an Italian (Salimbeni), and a Belgian (Bordet, who won a Nobel Prize for his research in his laboratory).[16] And obviously, the mere presence of someone can act as a powerful magnet for recruitment—even if they are not actively engaged in recruiting.

The importance of recruitment patterns becomes evident once one recognizes that each of the leaders had a particular bias, drafting individuals from his own educational network. Duclaux, a graduate of the prestigious École Normale Supérieure (ENS) himself, tended to recruit individuals from the *grandes écoles*, especially his own and the Institute of Agriculture, where he had a joint appointment during his years at the institute, as well as from his students at the University of Paris, Faculty of Science. In addition, Duclaux showed a decided preference for physical chemists. In contrast, Roux recruited medical doctors, especially with the aid of the chair in bacteriology at Val de Grâce (military medical school) and veterinarians from Alfort where another collaborator, Edmond Nocard, was located.[17] Roux tended to recruit from the less prestigious medical schools and preferred good technicians. Because Metch-nikoff was a Russian, the internationalization of Pasteur Institute was greatly facilitated. He emphasized zoologists.

Perhaps the most interesting aspect of their recruitment was that the bias was both intellectual and network oriented. In this manner, team diversity served to diminish the bias of any single leader and created the opportunity to have diverse research teams, teams diverse not only in biological specialties or medical specialties but in national cultures. This last point is worth emphasizing. Of the eight and one-half major breakthroughs mentioned earlier, four were made by foreigners (Metchnikoff, Levaditi, Bordet, and d'Herelle).

On the other hand, none of the people recruited by Duclaux were associated with major prizes. In Chapter 3, I noted that senior managers should avoid hiring stars, but that is what Duclaux did. He focused on graduates of the most prestigious schools. In particular, his choice of G. Bertrand, while understandable given his major discovery in biochemistry, was a particularly disastrous appointment. Bertrand actually abandoned the study of biochemistry less than ten years after he became the head of the new Institute of Biochemistry that had been created by Duclaux and focused on physical chemistry. He did not work in teams and instead treated his students simply as appendages to the laboratory. Since he controlled the major set of courses in biochemistry at the University of Paris, he effectively destroyed this area of research in France for the first three decades of the twentieth century. And within the Pasteur Institute, he became one of the major reasons why biochemistry was not linked to medical research or basic biological research. In short, Duclaux's personal characteristics illustrate how stovepipes develop and the dangers of having only one leader. We now have some of the personal characteristics that impact on how stovepipes emerge. One might ask why did not the leadership team intervene? The policy of the Pasteur Institute was research autonomy, and therefore, it was unthinkable that they would try to change the direction of his research, especially given his important discovery.

In contrast to the stars hired by Duclaux, individuals recruited by Roux and Metchnikoff from the less prestigious schools or foreign countries did quite well. For example, two of these breakthroughs were made by individuals trained in provincial medical schools (Nicolle and Delezenne). Calmette, who was trained in the military medical school, accounts for one-half of the major breakthroughs, since his prize-worthy research leading to the tuberculosis vaccine was performed both before and after the first thirty-year period that is the focus of our analysis of the Pasteur Institute. Finally, individuals trained at the University of Paris made two scientific breakthroughs, but both of the individuals had unusual career trajectories. Maurice Arthus spent most of his research career out of France, while Roux, before being trained at the Faculty of Medicine in Paris, had been expelled from the military school that trained Calmette because his unusual research challenged conventional wisdom. Thus, while the less prestigious tracks account for a minority of the researchers re-cruited, they do account for a majority of the major breakthroughs. As sug-gested in Chapter 3, it is preferable to recruit individuals who think "outside of the box" rather than those who do well in school, especially the most pres-tigious schools. Metchnikoff and Roux both violated the basic rule in France by not hiring the "best and the brightest" coming from what are called the *grandes écoles*, highly selective organizations with rigorous exams that place an emphasis on memorization rather than creativity.

The importance of Metchnikoff and Roux in encouraging cross-fertilization of ideas has been touched upon in Chapter 3. They were responsible for a number of practices (see Table 4.1) for increased understanding and for a re-duction in the decline of communication given differences in perspective within or between research teams. The lesser impact of Duclaux is partially explained by his heavy involvement in the political struggles of the Dreyfus Affair and in part because of his different recruitment patterns, which were previously discussed. However, as is apparent in Table 4.1, most of the mecha-nisms for encouraging cross-fertilization were present. It is worth listing these because *it is their combination* that became especially effective in encouraging

Table 4.1. Practices for stimulating cross-fertilization

Mechanisms for increasing understandability

1. Venues for training, rotation of young researchers, house organ

2. Duo research team leadership with diverse modes of thinking

3. Multicultural researchers

Mechanisms for reducing the rate of decline in communication

1. Elimination of status barriers

2. Creation of a knowledge research family

3. Emotionally supportive leadership

cross-fertilization between departments. Continuous learning was possible via the house organ for those who read it.

The combination of biases created a very open atmosphere that was the opposite of an elitist organization typical in France at that time. And this generated considerable cooperation between many, if not all, the divisions and departments within the institute.

The third organizational practice that helps overcome stovepipes is having revenue came from multiple sources. Indeed, generally the institute had surpluses in operating funds until the end of the First World War, which obviously reduces conflicts over resources. But several of the large bequests allowed for the creation of new departments and divisions and thus opportunities for researchers.

What were the sources of revenue? Ironically, despite Pasteur's emphasis on financial independence from the state, the Pasteur Institute began with very few operating funds outside of the state allocations because of appointments. Most of the money raised in the first public subscription after the discovery of the treatment for rabies was spent on the construction of the first building, leaving only 10,000 francs, which were used to establish an endowment, and nothing for the operating expenses needed in research. All but one of the five key researchers who formed the nucleus were employed in state educational organizations (the exception was Metchnikoff, who had private income). Furthermore, most of the operating funds largely came from the French state, which provided about 1,000 francs in a budget of 1,250 francs in 1890.

What gained the institute more economic independence was the development of an effective serum for diphtheria, which in turn led to a second public subscription that provided the initial funds for further expansion. This discovery led the state to purchase diphtheria serum with a grant of 2,000 francs in 1893.[18] Such an achievement was also due to the tireless work of Roux and his participation in a number of health commissions.[19] In general, the institute was able to maintain a quasi-monopoly on the production of diphtheria serum, despite the opposition of physicians and pharmacists, allowing it to cover the cost of applied research and continued improvement in the quality of vaccines and serums.

The absence of property rights for patents on drugs, as well as a general negative opinion of profit making in scientific institutions, led Pasteur to set an important precedent by donating his royalties to the institute. Pasteur had already started selling the animal vaccines that he had created before the famous rabies inoculation. In all, the royalties from these activities created new and stable sources of income.

Roux followed this precedent for both diphtheria and the tests he developed on tuberculosis.[20] Besides donating the potential profits from vaccines and serums, Roux went further and donated his prize of 1,000 francs to conduct research.[21] This motive of self-sacrifice, or science disinterested in profit, was continued when Laveran donated a substantial portion of his 1907 Nobel Prize money to create new research laboratories in tropical medicine.

More broadly, the publicity of the scientific breakthroughs, specifically shots for rabies and serum for diphtheria, along with the public subscriptions they generated, created an aura around the Pasteur Institute that led to a number of donations. Between 1887 and 1918, fifty-two bequests generated more than 50 million francs.[22] In particular, two sizable bequests allowed for the construction of a new building for the Institute of Biochemistry, as well as a new hospital for infectious diseases. The largest single donation was from Daniel Orsis, who left his entire fortune of 36 million francs; however, this did not impact on the budget until 1910, and then 40 percent of the revenue from his donation was placed in reserve. The diversity of sources reduced dependence upon the state and allowed the Pasteur Institute to expand rapidly into new areas of research, as indicated in the next section.

Again, we have in this case study a history lesson—the importance of multiple sources of funding—for innovation that has been replicated in various studies. The role of multiple funding sources for encouraging product or service innovation has been demonstrated in the literatures in research on welfare agencies and the creation of new knowledge specialties.[23] The best example of the effectiveness of a complex resource base is found in the history of universities in the United States, which have combined federal research funds with state funding and student fees as well as donations and bequests to build powerful organizations. Across time, universities have been highly innovative in developing new disciplines and research areas, although they have not necessarily avoided the problem of lack of cooperation between their colleges, where the stovepipe problem is more typical.

But how do we know that complex strategy, team leadership, and different sources of revenue are the three major organizational practices that explain how to integrate different divisions? Three pieces of evidence provide some confidence that in fact these are the important ones. The first set of facts to look at is the internal differences in the Pasteur Institute from 1889 to 1919. These three practices united the departments of medical technology, zoology, and production of serums and vaccines, but two other areas were not integrated: the Institute of Biomedicine as we have already seen and the department of tropical medicine. In both cases, stovepipes emerged because the leadership of these departments worked alone and not in teams and because the major researchers were graduates of the most prestigious schools of France. Where the departments and divisions were not integrated, no scientific breakthroughs occurred despite the fact that the two leaders had made major contributions. In other words, it was not a consequence of their lack of creativity but instead their leadership style.

The second set of facts to examine is the Pasteur Institute over time and also to assess what happens with changes in these organizational practices. For a variety of reasons, many stemming from the consequences of the First World War, the complex charter with multiple objectives disappeared. Only a single goal remained—public health. The members of the leadership team died and were not replaced, so that the team ceased to exist. The inflation reduced the reserves of the institute, bequests declined, and the institute had difficulty in matching salaries in the public sector of the universities. As a consequence, there was no new recruitment as well. As the Pasteur Institute aged, it split into more and more stovepipes with several notable exceptions where radical innovations occurred. A dramatic drop occurred in the number of scientific breakthroughs. The several exceptions were departments that were well funded and had diverse teams. Such exceptions prove the general rule about these organizational practices as well as indicate why the Pasteur Institute did not continue as it had previously.

The third set of facts involves making a comparison with other illustrative cases. The SEMATECH (SEmiconductor MAnufacturing TECHnology) case in Chapter 5 illustrates the same practices. At the end of this chapter, a number of historical examples are provided of at least two-man leadership or dynamic duos that were transformational in the context of the discussion of the need to think about transformational teams. The latter have been more common in the economic history of the United States than has previously been appreciated, and they have been associated with some of our most innovative organizations, demonstrating the advantages of diverse visions within the leadership team.

As noted in the beginning of this chapter, these three organizational practices not only help to overcome the problem of silos within organizations, but they provide a basis for changing the organizational context. We move now to Step Five.

CHANGING THE CONTEXT

Creating a Career in Biomedicine, Expanding Its Frontiers, Reforming Medical Research, and Solving Global Health Problems

If Step Four involves overcoming internal obstacles by creating integration, Step Five involves tackling obstacles in the larger external societal context.[24] The specific circumstances vary from case to case. In particular, the very special circumstances of French society and historical conditions that were obstacles for the Pasteur Institute would not apply in many other instances. At the same time, the general principles that this case illustrates are useful for those transformational teams interested in creating a new scientific or technological specialty (e.g., combining basic and applied specialties). This case illustrates the following four general principles:

1. Building a career for the new discipline
2. Expanding the frontiers by innovations in occupational specialties and research technologies
3. Changing the "know-how" of relevant researchers and professionals
4. Working on a global problem

These same organizational characteristics—complex charter, transformational team leadership, and multiple sources of revenue—provided the leadership with a vision and a power base that allowed it to transform the scientific and medical research environment or context.

Of these various changes, perhaps the most fundamental for establishing biomedicine as a new transdisciplinary research area was the creation of a career in this research area. But to do so within the context of France meant overcoming a whole series of blockages. Dividing a career into four stages—(1) scientific training, (2) initial research position, (3) publication outlets, and (4) promotion opportunities for senior positions—facilitates the identification of important patterns that affect whether researchers perceive that there are career opportunities as well as how they think about important problems. These stages also allow us to understand how reduced career opportunities at each stage can limit creative thinking. In each of these stages, an important set of rules blocked creative people in French science by selecting only a small minority on the basis of a competitive exam. This severity in the selection process combined with few opportunities for advancement led to considerable emphasis on not thinking "outside of the box" and a relatively homogenous elite of stars.[25]

But these were not the only lessons to be drawn; others involved the following topics:

- Rules about the kind of content for success, especially in the biological sciences
- Frequent need for a sponsor for selection, especially in medicine
- The length of the delay before access to a senior position in both areas

The argument is that these societal patterns prevalent in French society coupled with the lack of career opportunities prevented the development of new biological and medical specialties of research.

At the societal level, the economic and political dependency of public research organizations upon the state, the centralized control by the ministries of education and of health, and the restrictions on publication in the Academy of Science prevented the creation of new disciplines and therefore careers. The control by the French Ministry of Education on the creation of new posts in Faculties of Science, and by the Ministry of Health on the creation of new posts in Faculties of Medicine, effectively set limits on the possibility for new disciplines and thus for new areas of research to become institutionalized. But what aggravated this problem in France was the unusual Napoleonic requirement that a new discipline in a science faculty had to become part of the secondary program of instruction in the *lycée* (high school) and also be tied to an *agrégation* (special diploma for teaching). Obviously, this requirement considerably lessened the willingness of the Ministry of Education to create new disciplines in the university because it had such far-reaching systemic implications. Furthermore, the tradition was that once a new discipline was recognized, it was supposed to be established in all the provincial universities as well as in Paris simultaneously, although this was not the case with physiology. Again, the reverse side of this coin was that new initiatives from the provincial universities were not possible given the centralization of ministerial control.

So, the transformational team of the Pasteur Institute overcame these blockages by creating their own courses in biomedicine, the *grand cours,* which was mentioned in Chapter 3. They discouraged individuals from studying for the highly competitive exams and proceeded to promote individuals on the basis of their ideas. Finally, they recruited from a wide number of sources rather than relying solely upon the most prestigious, as we have seen. Thus, they created their *own venue* outside the power structure of France.

Leaders who create careers can be justifiably called transformational leaders. It is this problem-solving capacity for overcoming obstacles in the context of the organization that is the defining quality of transformational leadership. But establishing a career entails more than just creating a new course in biomedicine. To establish biomedicine as an important area of research, the Pasteur Institute had to create independent career lines and job opportunities to recruit creative people and further research in this area. To this end, the leadership team had to overcome three obstacles. First, there had to be growth in the number of positions available, that is, *opportunities*. Second, there had to be access to these positions relatively early in one's career and reasonable possibilities for *promotion*. Third, new publication outlets had to be created that would allow for research in this new discipline to be represented and for the researchers to receive *recognition* that would facilitate their promotion. Indeed, one of the more stunning kinds of institutional change that the Pasteur Institute accomplished was the way in which the creation of new journals seriously challenged the power of the Academy of Science.

New opportunities were created both within and without the institute, which experienced a rapid expansion as new institutes with the same name were created in various parts of the world. With diverse sources of funding, the Pasteur Institute grew rapidly in size from 1890 to 1915. In 1890, there were only five chiefs of services and several assistants, but by 1915 the institute had seventy-five full-time researchers.[26] This, however, was not the only source of job opportunities as related institutes were created elsewhere. Another key ingredient was the creation of a number of positions in different places that allowed one to pursue a career as a biomedical researcher, such as in Saigon (1891), Tunis (1893), Constantinople (1893), Nhatray in Indochina (1895), Dakar (1896), Tananarive in Madagascar (1898), Lille (1899), Brussels (1900), Brazzaville in the Belgium Congo (1908), Algiers (1909), Tangiers (1910), Chengdu in China (1911), and finally Bangkok in Thailand (1913).[27] The most obvious institutional change, one largely under the control of the Pasteur Institute, was the creation of a number of annex laboratories in Le Havre, Lille, Nancy, Lyon, Grenoble, Marseille, Montpellier, and Bordeaux. The positions made available by the institute and its sister institutes not only added to the quantity of opportunities, but also the variety. Because the latter were located in different parts of France's colonial world, as well as in the Mediterranean basin, they offered opportunities to do research on a wide range of different diseases. Typically, these offshoots started as production clinics where rabies shots could be administered and the serum produced, but many of them quickly evolved into full-fledged institutes that did public health research on topics appropriate for the specific country or region. In the first decade after the founding of the institute, when the operating budget was small, overseas posts became career ladders for creative people.

One appreciates better the meaning of these career opportunities if one compares this rate of expansion with what was happening elsewhere. In the Faculty of Medicine in Paris, another main site for medical research, only four new chairs were created between 1880 and 1914. In the interwar period, there was a greater expansion with sixteen new chairs. But of these twenty new chairs during the sixty-year period, thirteen were clinical chairs and therefore little concerned with basic research.[28] In the Faculty of Science in Paris from 1891 to 1913, four new chairs were added to the existing six, but most of these were in the traditional areas of biological research such as botany, zoology, and physiology.[29]

Perhaps one of the most important ways of recruiting creative minds and facilitating radical innovation is to promote individuals rapidly. Such was the case at the institute, where many researchers were quickly appointed *chefs de travaux* (project leaders), then advanced to chiefs of laboratories and/or chiefs of departments at early ages. This rapid expansion at the bottom coupled with rapid promotions to higher positions—some individuals became chiefs of laboratories within a few years—attracted a large number of interested individuals looking for a career. It also meant that they were receiving decent salaries that allowed them to have a normal family life. And perhaps most critically, it allowed for fresh ideas to be heard and investigated. By 1910, there were already thirty-seven chiefs of service and another eight department heads in the Pasteur Institute, and this number did not include their counterparts at the various institutes overseas. If one remembers that the full-time researchers numbered about seventy-five at the institute in 1915, then more than one-half of them had their own laboratories, a situation considerably different from the one in either French science or medicine of the time.

What is especially noteworthy, particularly in the context of limited mobility in the traditional institutions of France, was the young age of many of these individuals. Borrel, Martin, Mesnil, Levaditi, Besredka, Delezenne, Jacques

Duclaux, Charles Dopter, and Edouard Dujardin-Beaumetz were all in their early or mid-thirties when they became laboratory chiefs, while Maurice Nicolle, Salimbeni, Weinberg, Mouton, and Marie were barely forty when they reached this echelon. Also, Weinberg, Chatton, Pozerski, Abt, and Berthelot had become assistants, the echelon just below chief of a laboratory, in their early thirties. This meant that the Pasteur Institute was *a very young organization* in its effective leadership at the research lab level at the turn of the century. The same pattern was repeated for the researchers who went overseas, especially those who founded various institutes.

A comparison with the prevailing norms in the Faculty of Science in Paris again provides some indication of how exceptional these early promotions were. The average age for promotion to a chair in zoology or botany was 51.7 years.[30] In the Collège de France, it was slightly younger during the same period and 48 years in the medical faculty of Paris.[31]

But besides job opportunities and early promotions, researchers needed outlets in which to publish the results of their research. The Academy of Science was controlled by a few biologists who thought only in conventional terms about fauna and flora and were not interested in the new scientific theories associated with Pasteur's work in bacteriology.[32] Because the Academy controlled *Compte-Rendu*, the official outlet for major research findings, the biologists could block publication of papers in this area. As I have already mentioned, Duclaux perceived the need to create a new journal, *Annales de l'Institut Pasteur,* to provide an outlet for the publications in the new science of bacteriology. Thus, the creation of this journal broke the monopoly of the Academy of Science and furthered the careers of the new and young researchers at the institute. Another journal with a strong applied orientation, *Les Annales de la Brasserie et de la Distillerie,* was started by the institute in 1901.

Creating the career for biomedical researchers, not only in France but worldwide, was not the only major societal change. The transformational team also expanded the knowledge base of biomedicine. A theme of the Introduction is how fast knowledge is growing and in the process changing. A century earlier in the new research area of bacteriology, the same was true. There was a fierce competition between the Pasteur Institute and its counterpart in Berlin with major discoveries being made every few years. But what is striking about the leadership team of the Pasteur Institute is how much they pioneered in the expansion of the knowledge base of biomedical research. What needs to be stressed is their creation of new occupational specialties, technologies of medical research, and of course, areas of biomedical research. These represent fundamental changes in the context of French science but also, in several cases, for the entire world, and they reflect one of the most important ways in which transformational leaders need to think about the world of science.

The institutionalization of a new specialty is by no means a simple affair. Many steps had to be taken. Again, one of the most unusual innovations was the creation of the first institute for the study of biochemistry anywhere in the world. This was the result of the vision of Duclaux, who had been advocating this approach to the study of microbiology since 1882, and of receiving a large grant to build a second building for research in this area. Later he also added several distinct laboratories in biophysics. Meanwhile, Duclaux expanded the applied research side of biomedical research to include the applied fields of industrial and agricultural research. During the first few years of the institute, Duclaux kept proposing these additions but Pasteur argued against them, preferring a singular focus on biomedical research. It is interesting to observe that once Pasteur died, the board of directors immediately implemented Duclaux's proposals.

Metchnikoff added another paradigm to the study of immunology with his cellular approach to immunology. Interestingly, competition existed between France and Germany and their two quite different paradigms, one cellular and one blood-based, and both were recognized in 1908 with a Nobel Prize. But beyond this, the research in Metchnikoff's lab also led to another Nobel Prize in this area, where a large number of papers were produced across the several decades before the end of the First World War.

Metchnikoff observed the new techniques developed by August Wasserman in research on blood and sent Levaditi to London to learn these so that the Pasteur Institute could begin research on syphilis, which it did. Indeed, this was the last major project on which Roux and Metchnikoff worked together. Roux used his prize money from the Copley Medal to establish an animal laboratory of monkeys on which the research was conducted.

Roux's vision was also at work with the addition of three new specialties to the institute: protozoology, biology of radium treatments for cancer, and chemical therapy. In the case of protozoology, he hired Laveran when he retired from the military where he was frustrated in doing research in tropical medicine, despite his pioneering research in the causes of malaria. Roux gave him his own laboratory, and later when Laveran won a Nobel Prize for his research, he donated part of the money to establish a broad research program in protozoology with three distinct laboratories.

The idea of radium treatments for cancer was precipitated by the pioneering research of Madame Curie and, of course, the establishment of a center of research at the University of Paris. Roux seized upon this opportunity and supported a new division of research on the biology of radium treatments. But perhaps the most interesting appointment and the creation of a new area of work was chemical therapy. In 1911, he hired Ernest Fourneau, a pharmacist, but one who had been trained in Germany and came with a whole new theoretical approach to the study of therapies for patients. In this laboratory, sulfa drugs were developed but not patented by the institute so that it earned no money. His laboratory later produced another Nobel Prize for Bovet, the Italian, who discovered antihistamines.

As I have already noted, Roux had a special interest in new technologies for research in biomedicine, one reason he was interested in both the biology of radium treatment and chemical therapy. His first efforts in this area were the development of the blood serum from horses for treating diphtheria. Afterward, Roux convinced Pasteur that there was a need for a hospital to conduct research on diseases and that it had to be constructed in a special way to ensure the quarantine of the individuals with infectious disease. The hospital, which was designed by Roux himself, was the first in Europe to isolate patients during treatment.

Finally, new areas of research that emerged from work in the Pasteur Institute were also established. The institute is not given credit for it, but Borrel discovered viruses in the first decade of the last century and began to work in virology. Unfortunately, he left to teach at a university and this research program died. The same did not occur with the discovery of phage by d'Herelle. Although he also left, the phage research program continued across the years and eventually led to some of the major work recognized by Nobel Prizes in the 1960s.

Together these additions to the knowledge base represented a rapid expansion in the number of specialties within biomedicine and reflected the importance of leaders who continually monitor how the science was changing. What is distinctive about the institute is the number of firsts that they were associated with in the beginning of the development of this new specialty of biomedicine. And perhaps it is worth repeating how quickly knowledge evolves and how essential it is for leaders to recognize this.

So, establishing a career in biomedical research was a key element in the strategy of reforming both biological research in the universities and medical research in the hospitals of France. But it was not the only component. The other was teaching practicing physicians experimental medical research and new ways of thinking via the new biomedicine course, which was previously described. What is not apparent in the simple description of the addition of this course is that it represented a kind of coup d'état in the context of French scientific and medical education. New courses, and especially one as ambitious as this one, would have to be approved by both the Ministry of Education and the Ministry of Health because it involved biomedicine, the combination of biology and medicine. The Pasteur Institute, however, ignored both ministries and created another first in the world. Furthermore, the course was highly successful, training about one hundred physicians per year including foreign doctors. The course was revolutionary within the context of French, and we might add world, academics of the time, not only because it combined two disciplines but because it was taught by twenty-three individuals, most of them renowned in their own research area.

In the first thirty years of its existence, the Pasteur Institute made a number of contributions to world health, starting with the cure for rabies (the diverse research team of Pasteur) and continuing through the serum for diphtheria (two diverse teams of Roux), the solution for typhus (the diverse team of Charles Nicolle), the attempts to solve the problem of the plague (Yersin and others), and the vaccine for tuberculosis (Calmette and Guerin's broadscope team). These contributions exceed those of any other research institute in medicine or bacteriology of the time. Nor do these contributions consider the many other health problems that the institute studied and, while making some progress, was not able to eradicate.

One of the most remarkable aspects of the health agenda of the Pasteur Institute was its *global reach*.[33] Rapid deployment teams were sent to diseases around the world, not just in the French colonies. Pasteur started this practice by sending Nocard, Roux, and Louis Thuiller to Cairo to study cholera during an outbreak in 1883, where the latter died from the illness. This practice was continued throughout the next three decades. For example, Adrian Loir was sent to Australia to identify whether a bacteriological method could be developed to control the explosion of rabbits in 1887, the year before the institute was opened. Yersin went to Hong Kong (plague) in 1894; Paul-Louis Simon to India for the same reason in 1896–98; Calmette and Salimbeni to Porto to test a sero-therapy for bubonic plague in 1899; Marchoux, Simon, and Salimbeni to Rio de Janeiro to study yellow fever during an outbreak in 1901–5; and Martin, Axen Leboeuf, and Emile Roubaud to the Congo for sleeping sickness in 1906–8, among others. Such a global vision of health problems can only be admired.

Earlier in this chapter, I observed that many job opportunities were created with thirteen sister institutes, eight within the colonies and five outside. Although some of the motivation for this expansion was a reflection of the French view of empire, another part of the motivation was to directly attack local health problems in situ. Interestingly, the Pasteur Institute had become a global organization long before manufacturing firms became international.

What were some of the local health problems that these sister institutes examined? Calmette went to Saigon to 1891 to establish the first one and while there began to work on a serum for cobra snakebites. Besides the plague, Yersin established a branch of this sister institute to work on the agriculture of rubber trees, which was a very important crop for French Indochina. Charles Nicolle won a Nobel Prize for his work on typhus in Tunisia, a major illness in that country. Many other, but less famous, examples could be provided, but the key point is that working on these issues benefited the world in general.

In summary, transformational leaders when building a new transdisciplinary specialty need to concentrate on how they can create the necessary competencies, which is not just a question of scientists but includes technicians, and I might add, in each of the different arenas of the idea innovation network. Each society will have different blockages relative to the availability of the necessary skills sets and in the quantity that are needed. In Chapter 6, we return to this problem in the United States and, in particular, the need for technicians of all kinds. Retraining existing scientists and professionals relevant to the particular new specialty should be part of the vision. In addition, there is frequently a need to create new subspecialties within the new transdisciplinary specialty as well as a variety of new technologies. Focusing on solving a national and/or global problem provides the emotional cement that will encourage people to work together.

EVOLUTION, TRANSFORMATIONAL TEAMS, AND THE LEARNING ORGANIZATION

The two themes of this chapter—the organizational practices to integrate divisions within an organization and strategies to change the organizational context—can be easily integrated with those in Chapters 2 and 3. These practices provide additional insights about open innovation, absorptive capacity, and the role of emotions in learning. In particular, they elaborate on how one can construct a learning organization by continuing the theme of stimulating cross-fertilization at the level of the organization, both internally and externally.

These three organizational practices provide new ways of thinking about leadership, especially the emphasis on a team, and how organizations should relate to their context in attempting to change it. The evolution in how radical innovations are created requires more complex leadership models such as a team and more aggressive ways of interacting with the environment. The idea of transformational leadership in this context bears some similarities to the concept in the management literature.[34] These leaders had idealized influence and provided inspirational motivation and intellectual stimulation while focusing on the individual. Duclaux, Metchnikoff, and Roux varied in these particular qualities, but together they provided these four attributes that have been used to define *transformational leadership*. However, a major difference in the usage of the term in the management literature and in the context of this chapter is that in the former, the emphasis is on the relationship between the leader and the researcher. In this chapter, *transformational* means not what the leader is doing to the researchers—transforming them—but to the environment, changing it. The vision of improving French public health and solving global medical problems provided the inspirational motivation and, of course, the researchers remarked on the intellectual stimulation especially from Roux and Metchnikoff. *In restoring the innovation edge, the key is how transformational leadership teams are changing the larger society.*

In some respects the idea of a leadership team is already present in the management literature. As is well known, a distinction is usually made between the "inside man," who is usually the chief executive officer, and the "outside man," who is typically the president. This recognizes that there are essentially two tasks to combine: (1) the internal integration of the organization, which for a research organization means creating the coordination of different departments and encouraging the cross-fertilization of ideas between them; and (2) the external integration of the organization, which for the research organization means monitoring competitive challenges in the marketplace, including scientific breakthroughs and technical advances that are relevant. These two tasks in themselves point to the advantages of having a leadership team.

But I would suggest that rather than divide the two tasks in two separate offices, it might be better for both of the two leaders to concern themselves with internal and external integration. There is an obvious relationship between them, as is indicated in the case study just presented, because each leader is likely to think about the corresponding set of issues in disparate ways.

With historical hindsight one begins to see how important team leadership has been in some the most successful companies in the economic history of the United States and elsewhere. Consider the team of Sloan and Kettering at General Motors, which created the divisional model capable of producing cars for different social categories and coupled it with a policy of major technological innovations that made GM the dominant company in the world during the 1950s, before the change in leadership described in the Introduction. The key point about this duo is that without the support of Sloan, Kettering could not have pioneered so many radical innovations. After this, the dynamic duo of Eiji Toyoda and Taiichi Ohno at Toyota created the learning and innovative supply chain (see Chapter 5). Again, the emphasis on productivity has led everyone to discuss just-in-time delivery, which is *not* its major achievement. The importance of the way in which the supply network for Toyota was constructed is that it allowed Toyota to continually make technological advances.

But given the various complexities and challenges created by the evolution in science and technology, leaders of business and research organizations can no longer simply regard the competitive situation; they must monitor how various new research areas relevant to their products and services are emerging and where the training of individuals that have these competencies occurs. But even more than that, the case study of the Pasteur Institute indicates how they should be actively involved in building the new human capital not only for their firm but for their sector as well. The four principles outlined add a number of new ideas to the learning model of organizations that was introduced at the end of Chapter 3.

5 CONNECTING THE RESEARCH ARENAS IN THE IDEA INNOVATION NETWORK
Step Six in Restoring the Innovative Edge

THE SINGLE BIGGEST obstacle to restoring the innovative edge is probably the lack of frequent and intense interaction between the differentiated arenas in a number of high-tech sectors. The gaps result in the slow transmittal of scientific discoveries or technological advances, creating what is called the *valley of death*. Without integration of all the six arenas, radical product and process innovation becomes more difficult. Again, this continues a theme of the last several chapters about the importance of cross-fertilization of ideas but now at the sector level. The remedy would appear to be simple: construct linkages to close the gaps between the differentiated arenas in the idea innovation network.

But this remedy is not simple for a variety of reasons. First, the valley of death has emerged from a variety of causes, most notably the presence of silos or stovepipes that we have discussed in Chapter 4. Additional examples of this are provided in this chapter. It is not easy to overcome this problem as the example of the Pasteur Institute should have demonstrated; it requires a lot of effort. Second, it is not just a single idea innovation network that needs integration but a set of them attached to the supply chain and especially in complex products. For each technologically sophisticated component in the supply chain, there is an idea innovation network, hence the term *technical systems,* which has recently emerged to capture this reality. Illustrations of the importance of technical systems are provided in the opening of this chapter and discussed further in sections two and three. Third, precisely because there is so much variation between sectors in the nature of their idea innovation networks (number) and supply chains (extensiveness), there is not a single policy recommendation that can be provided to all of them. Therefore, this chapter considers four prototypical kinds of connections with one or more examples and ends with a typology for describing some of the general patterns that could potentially help policy makers craft the appropriate kinds of policies for specific types of industrial sectors.

Let us begin with two relatively simple products and demonstrate how different radical innovations can be for the success of the country and its trade balances, depending upon who supplies the various components in the product. The first case is Apple's iPod. Even though I have cited the car and the airplane as extreme examples of products with many components, the iPod has hundreds of parts, some of which have considerable technical sophistication that confers higher profits on the supplier.[1] In the 30 GB 5th-generation iPod (video iPod) of 2005, Toshiba of Japan provided the hard drive. Interestingly enough, Toshiba had to keep its price low on the hard drive because it could be replaced with a flash memory. A joint venture of Toshiba and Matsushita produced the display assembly, a technology that the United States had invented but had lost control of because we did not keep improving the product. Two U.S. companies manfuctured particularly important microchips, Broadcom and PortalPlayer; the former produced a chip that controls the video playback, and the latter one that manages the functions. This latter company participated in the design but was later replaced by another supplier as the pace of technological change in this market accelerated. This analysis could not identify the producers of the lithium-ion batteries, but they were probably Japanese since three companies in Japan control a large percentage of this component market. An additional three memory chips were manufactured by Samsung, a South Korean company. This study did not isolate the suppliers of the more than several hundred other parts except to note that mechanical parts that allowed for the small size and some microchips probably had a high technical sophistication. Apple outsourced the assembly to a Taiwanese firm, Inventec Appliances, who owns factories in China where the product was constructed. The case illustrate three important points: the supply chain is global and involves multiple countries; even a small high-tech product can be complex; and quick changes in the technology can shift who supplies a particular component.

Profit-wise, Apple was estimated to make about 36 percent of the retail price and even more in its expanding base of Apple stores. Part of the profits in this example acrue to Apple because of the difficulty that competitors have in matching the secure downloading capabilities of iTunes and its very wide access to music. In other words, Apple is not only controlling the design and innovation of the iPod but has behind it some major advantages in the choice of an application that is quite popular. Finally, as is well known, these small devices have profilerated the number of applications, illustrating the principle of multifunctionality.

The success of Apple illustrates many of the ideas suggested about the advantages of pursuing a strategy of radical innovations, moving from iPhone to iTunes to iPod and now iPad. The design, ease of use, multifunctionality, and technological sophistication appeal to the modern consumer not only in the United States but worldwide. It is estimated that 40 percent of the sales of iPod are outside this country. Although the iPod is a success and a perfect illustration of a global network, one where the innovation is controlled in the United States, it is also an example of a failure because so few components are produced in the United States. The large profits of Apple do not provide enough jobs to replace the number of workers who would have been employed in this country if more of the components, especially the technologically sophisticated ones, had been produced in the United States. It underlines that policy has to be concerned not only about the production of sophisticated consumer products such as the iPod but more critically the various components that are in the supply chains. This is another reason for our negative trade balances despite the seeming success of Apple and companies like it. We have failed to invest in increasing the technological sophistication of the components, most notably that of the display, which was invented in this country. Furthermore, Apple remains vulnerable to the possibility that some Taiwan company or perhaps Nokia may produce the next generation by working closely with the various suppliers of Apple's components to develop the next generation.

To take another example, consider the recent arrival of notebooks. It has even more major components. For example, in 2005 the Hewlett-Packard (HP) nc6230 Notebook PC had the following major components that were purchased in the following countries: main chip and WiFi (Intel, United States); Ethernet controller (Broadcom, United States); cardbus controller (Texas Instruments, United States); I/O controller (SMSC, United States); display assembly (Toshiba/Matsushita, Japan); Windows XP (Microsoft, United States); 60 GB hard drive (Fujisu, Japan); DVD-ROM drive (Matsushita, Japan); 512 MB memory board (Samsung, South Korea); DDR SDRAM memory (Hynix Semiconductor, South Korea); and a battery pack probably from Japan, plus other parts.[2] Because more of the components of this product are produced in the United States, the success of this particular brand of notebook translates into more jobs for American workers.

The key point about both examples is that it is not just radical innovation in the end product but also radical innovation in the various components that is necessary to create employment in this country. Not only must the idea innovation network be coordinated for the end product, but it is also necessary to consider the various other networks attached to the components in the supply chain. These two examples illustrate not only the idea of technical systems but also that particular components, such as the display and the battery, play an important role in multiple supply chains.

By cooperating with other organizations, gaps in the idea innovation network can be closed, stimulating the rate of radical innovations. What is the evidence for this? Using the one hundred awards provided by *R&D Magazine* annually, one can plot the trend of those awards given to two or more organizations. As observed in the Introduction, there are three limitations with this data, the absence of awards in the defense industry, computers, and pharmaceuticals and the constant rate of one hundred awards regardless of the number of radical product or process innovations discovered in a specific year. But given these limitations, the awards demonstrate the advantages of network cooperation and coordination. Already fifteen awards out of a total ninety-seven (subtracting out foreign awards) were given to cooperating organizations in 1971. By 1984, the number had risen to thirty-eight out of eighty-six national awards. By 1997, the number rose to sixty-seven out of eighty-eight, a high that was repeated in 2006.[3] In the intervening years, the number fluctuated around fifty, or more than half. The message is clear that interorganizational collaboration and cooperation is necessary to achieve radical innovations, at least those that are commercialized. This research also provides additional support for the evolution of the idea innovation network because there has been a steady increase in the proportion of winners that have cooperated.

Because the evolution toward greater product complexity is such a powerful force, the first section of this chapter focuses on two examples of this evolution pattern. The first example is universities. More and more cooperation will be necessary between different research organizations within the *same* arena, such as universities or universities with the national research laboratories in basic research or between firms in manufacturing research. The reason why this cooperation is necessary is the growing difficulty of research problems. Despite the name, even universities cannot have all the competencies required. This example involves information technologies (IT), and it illustrates again stovepipes and their causes. The second example is that of IBM and computers. It illustrates the strategic problems that the growing demands for customization pose for large firms that have to develop multiple relationships with different customers.

The examples of the Apple iPod and the HP Notebook demonstate how important the supply chain is. Therefore, the second section focuses on how to make the supply chain a learning and innovative chain. The specific case is the automobile sector, which while not as high-tech as IT or computers still spends more than 4 percent of is gross revenues on product development and innovation, close to the 5 percent level that officially defines the high-tech sectors.

As noted in the Introduction, the automobile industry is the icon of the past, but the icon of the New Economy is the semiconductor industry, which is a platform technology for most electronic products. There is a major difference between a firm that turns its supply network into a learning network to handle technological change and a research consortium that unites most of the companies in the industry with the objective of making them technological leaders. The specific example is the research consortium SEMATECH. So, the third section of this chapter reports how a transformational team and an interorganizational network saved the American semiconductor industry and how it has evolved across time to be more than just a national research consortium. Research consortia, if managed properly—which is why this case is so interesting—offer both hope and principles for restoring the innovative edge to the United States.

In the fourth section, a special kind of connection or cooperation is discussed, creating spatial opportunities for learning and innovation following the pattern of the success of Silicon Valley. Many governments, including our own, want to duplicate this pattern, and therefore it is worth considering what the special circumstances are for establishing a learning *region* that combines sectors. The key lessons for creating Silicon Valleys are establishing practices that reduce risk and encourage collaboration. Several examples outside the United States are provided as evidence that these two lessons are the correct ones to draw from Silicon Valley rather than other interpretations that are abundant in the literature.

The first four sections illustrate a specific kind of connection in the idea innovation network:

- Filling gaps created by the evolution toward greater complexity
- Creating learning supply networks to handle technological change
- Creating research consortia to turn an industry into a technological leader
- Creating regional spaces for learning and innovation

In the Introduction, I stressed the importance of recognizing that each sector has its own idea innovation network and, one might add, supply chain because the investments in research and development (R&D) vary greatly both nationally and globally. Therefore, each has its own discrete problems, likely evolutionary failures and distinctive gaps between diverse research teams located in various arenas. In the fifth section, I propose a simple typology for describing some of the differences as an aid for managers and policy makers to consider when fine-tuning industrial policy, which will be the topic of Chapter 6. These differences in problems, failures, and learning gaps across sectors can only be partially addressed in this chapter. I say *partially* because it would require a whole other book to provide a detailed analysis of each of the sectors, both low-tech and high-tech, in the American economy as well as the various noneconomic sectors (such as health, education, welfare, defense) that need scientific breakthroughs and radical innovations. Given space limitations, I will discuss four generic types. In addition, the extended examples in this chapter include major high- and medium-tech sectors—automobiles, computers, information technology (IT), and semiconductors, which provides a considerable amount of evidence for Step Six in restoring the innovative edge.

FIRST DEGREE OF CONNECTION

Filling the Learning Gaps Created by the Evolution
Toward More Complexity

Detecting where there is an absence of a connection or a weak connection requires a diagnosis upon the part of both government and business leaders in their specific sector, because each sector, as it evolves, develops its own unique problems. Given the many potential gaps in networks—165 for just the eleven high-tech sectors listed in the Introduction—as well as the potential underfunding of certain research arenas in any of the six arenas in each of these eleven, the discussion of failures can consider only a few prototypical cases. Two parts of the idea innovation network will be emphasized as examples of the kinds of problems created by growth in the difficulty of the research problem. The first case is the problem of cooperation in basic research, in this instance, between American universities in developing diverse teams that have the necessary expertise for interdisciplinary problems. The second example moves to the opposite end of the idea innovation network and handles the complexity created by the divergences in taste and growing demands for customization discussed in Chapter 1. In the past, large business firms with their closed innovation model were able to consider how best to solve the problems of commercialization, but this model is no longer viable, as we have seen. The dominant business firm has to develop networks to organizations that are expert in providing services to customers. Each of these cases speaks to contemporary problems involved in our innovation crisis and provides models of what can be done to solve them.

One of the most important engines of scientific breakthroughs is the university. But despite the name implying that all areas of science and technology are contained within the same organization, the reality is that increasingly each university can only cover some competencies in each of the rapidly growing scientific and engineering areas of research, to say little about the new disciplines that have emerged in the last twenty years such as computer sciences, material science, and in particular nanotechnology, and most recently, alternative energy sciences. For complex problems in basic science and engineering, increasingly universities themselves have to forge diverse research teams across their organizational boundaries, and when they do, they discover a whole series of obstacles that prevent effective communication and innovative outcomes.

Because various funding agencies are encouraging more multiple-discipline collaboration and because the complexity of research problems grows over time, the issue of how universities can cooperate more effectively is an important one. For example, the National Science Foundation (NSF) funded a program called Knowledge and Distributed Intelligence within IT that was designed to generate, model, and represent more complex and cross-disciplinary scientific data.[4] Although this is a case study of sixty-two projects funded in a single NSF multidisciplinary program, it provides a cautionary tale about the problems of interuniversity collaborations, even when there are no proprietary issues.[5] The necessity of having multiple universities involved was established because the greater the number of disciplines involved in the project, the greater the number of universities that participated.[6] The following are the percentages of disciplines that were involved: computer sciences (16 percent), electrical engineering (13 percent), other engineering and psychology (12 percent each), physics and mathematics (9 percent each), and biology (8 percent). The first irony is that this project was designed to produce more effective information technologies but also demonstrated that the various electronic means of communication were not effective, an argument that I have made previously. Although e-mail was used extensively, it did not appear to solve the coordination problems represented by the dispersion of disciplines at different universities.[7]

The second irony is that precisely when *more* effort should be applied to overcome the double problems of disciplinary and spatial distance, actually less effort was used. In other words, the greater the number of universities involved in the multidisciplinary project, the lower was the level of faculty, postdoctoral, and graduate student supervision or face-to-face contact and, most surprisingly, much lower was the level of monthly project meetings. The pattern of coordination shifted as well to a conference or workshop, which is less effective than the face-to-face interaction that encourages spontaneous problem solving and passage of tacit knowledge.[8] This drop in coordination had a price; it reduced the amount of innovation measured in multiple ways consistent with the argument in Chapters 2 and 3. Diverse research teams can produce more innovation, as this study demonstrates, but it requires that more effort be made to coordinate across paradigmatic and spatial distances for these objectives to be achieved in multi-university projects.

The litany of problems that the multi-university projects encountered provides a number of insights about why interorganizational collaborations are difficult to achieve. The distinct organizational blockages of American universities that explain this phenomena are the differences between semester and quarter teaching schedules; the procedures for budgets and especially subcontracts; contract languages, in particular relative to patent rights; and the compatibility problems of computers and software. Although many of the principle investigators (PIs) started with a great deal of enthusiasm, this diminished over time as the frequent meetings between the campuses became more and more difficult to maintain.

What is a potential solution? Because face-to-face communication is so effective in stimulating innovation, it logically follows that some form of colocation is desirable. But given the problem of expensive equipment not being very movable, what needs to be colocated are the PIs for long and frequent meetings to stimulate the cross-fertilization that is needed for radical innovations.

To do this requires some advanced planning not only by the PIs but also their respective universities. Housing should be set aside, including housing for families during summer vacation and winter breaks. Only one-fifth of the PIs used their sabbatical for the purpose of working together in the same place. *But using their sabbatical for this purpose should not be necessary; NSF's lack of support for colocation in its policies is one of the major reasons why this agency is not achieving more innovation from its investment in multidisciplinary research.* Its policy is *not* to fund PIs except for two summer months. This policy makes collaborations much more difficult than they need be. If they changed their policy to fund one semester each year for one of the PIs and the PIs alternated locations in different semesters, many of the coordination problems would be diminished. Given the recent large increase in the NSF budget, this alternative is now a distinct possibility and is one of those exceptions where money is part of the problem, contrary to the general argument in this book. The discussion of the SEMATECH project in the third section illustrates how important *colocation* is.

IBM provides a classic example of a firm that has to change its business strategy when the dominant design changes. Several case studies that detail how IBM reinvented itself illustrate a number of the points that have been made.[9] Consistent with the idea that in liberal market economies firms worry about their intellectual property, IBM pursued the development of products in its own research laboratory and covered its ideas with an enormous number of patents. It refused to cross-license except in a few instances. This policy led to the hugely successful mainframe 360 computer, which became the dominant design for the computer industry. When IBM needed expertise, it bought it in the form of equity investments in companies.

Just as Xerox PARC did not exploit many of its patents effectively, the same was true for IBM. For example, IBM developed robots and automated production systems that it never commercialized. Sometimes it purchased patents to slow down the rate of innovation by competitors as with the electric typewriter.[10] Many of the problems IBM faced occurred because of the lack of integration between the basic and applied research conducted in the research centers and the manufacturing end of the business, another typical example of path dependency in the United States—in this instance, industrial research departments tended to be stand-alones. This analysis illustrates how the idea innovation network theory can highlight blockages within firms, that is, the failure to develop internal networks. As a consequence, IBM had problems with its silicon semiconductors and in successfully exploiting some of its software achievements as well as its personal computer (PC).[11]

A number of scientific and technological changes occurred that made the dominant design of the mainframe computer obsolete. The growth of computer science departments meant the science advanced quickly and led to the development of new kinds of products such as the minicomputers. The constant improvements in processing speed, memory, and the like resulted in the continual downsizing of computers. A number of small companies emerged to provide different kinds of products and software. And of course, the Internet changed dramatically the competitive world of many of IBM's customers. In fact, many of the same changes that affected the telecommunications industry also influenced the computer industry.

As was suggested above, one of the pressures toward network solutions to product innovation is the increasing emphasis on speed. IBM discovered that its own research laboratories could not develop products fast enough to be competitive. One of the reasons for this failure was the lack of integration between the different research arenas. To solve this problem, the company created a new kind of joint research program to stimulate greater speed. Again, it is worth emphasizing that by working directly with the customer, one can develop products faster despite having to cross organizational boundaries.[12] But this new program did not prevent the company from suffering a huge deficit, which in turn forced more radical changes in strategy.

One consequence of this deficit was the hiring of a new CEO. Louis Gerstner, the CEO who came in 1993, recognized that there was much more to be gained from developing customized products and services for the customers than from concentrating only on the hardware of computing.[13] As IBM began to do research with its customers to provide unique solutions for them, it recognized that it also had to shift from closed technology to open technology, in part to gain more customers. It began to sell its products to its competitors to spread its particular solutions in the marketplace. Internally, this also meant changing the patterns of researcher recruitment and the construction of more diverse research teams that combined more arenas of research.

IBM dramatically shifted at the same time its network of alliances, changing from an emphasis on equity to one of nonequity with many new partners that were in different technologies—in contrast to the previous policy of usually having equity partners in the same technology.[14] A key point is that as one shifts from equity to nonequity alliances, the strategy also changes from one of exploitation or control to that of learning and cooperation. With coopera-tive relationships, more learning occurs and both partners grow, a point that is stressed in the next section. The exploration alliances provided many op-portunities for IBM to learn and to supplement what it was learning while providing customized products and services to its customers.

As is well known, the company is again quite successful. IBM's open archi-tecture and learning alliances led to a considerable growth in the business of customized services. By 1995, IBM had revenues of $12.7 billion in this division and by 2007, the revenue represented 55 percent of IBM's total world income.[15]

In summary, IBM solved the various problems associated with an economy that sometimes places too much emphasis on competition; it abandoned its concerns over property rights and began working with many small companies to develop new kinds of products and services, using a network approach to continue its learning. But it was a crisis that forced the company to move in new directions. Also, the case study of IBM offers another one of those success stories that on close examination is less of a success that it might appear at first glance. Now that 71 percent of its employees are located outside the United States and the company is expanding its research labs outside the country, this success is not translating into more jobs for Americans.[16]

SECOND DEGREE OF CONNECTION

Creating Learning Supply Networks to Handle Technological Evolution

The current trend is to design complex products as components that are as-sembled, as illustrated in the examples of Apple and HP. As products become more complex in terms of both the number of functions and the ranges in which these are offered to meet customized tastes, the supply chain becomes more and more central not only as a source of the components but, more criti-cally, as multiple opportunities for learning and for developing more complex and radical innovations. The supply chain is the potential site for both manu-facturing and quality research, including research designed to reduce risks to the environment. The extremes in complex products are cars, aircraft, and space vehicles.

Just as the idea innovation network has six arenas, supply chains have mul-tiple tiers: technical systems, components, and materials. In the Japanese sup-ply chain, an example is provided of the first-tier component of seats, followed by the second tier of various parts, and so on. The motor is a complex technical system. In addition, supply chains involve procuring the manufacturing equipment and providing various services. Because complex products are assembled from components, it is theoretically possible for a radically new product with considerable advances in technology to be based largely on one component (e.g., a hybrid engine in cars). But even this radical advance in energy conservation required reconsidering a number of other aspects in the design of the car and, by implication, major advances in the other components. More typically, a radically new product requires radical advances in most, if not all, of the components. Thus the 30 percent reduction in fuel consumption in the Dreamliner required technical advances—and major ones at that—in most components, making the learning and innovation in the supply chain critical.

When the components are produced internally, as in the case of GM or Boeing during the 1980s, the opportunities for learning are likely but not necessarily reduced. Even when the firms are external, learning may be blocked depending upon how the dominant company treats the suppliers.

The automobile industry, like other medium- or high-tech industries, has been experiencing continual technological evolution over the past three decades. The processes of increasing complexity—air bags, analytic converters, chips that monitor performance, global positioning system (GPS), skid controls, and so on—in the automobile and the differentiation of tastes have put more and more stress on the supply chain to keep technologically abreast and to develop new products rapidly. Comparing automobile industries in different countries provides a good illustration of how supply chains can be made into learning chains. The best example of how supply chains can be made into sources of innovation is how the Japanese manage their automobile supply chain. It is worth an extended discussion in contrast with how GM managed theirs.[17] The construction of the product development team in the Japanese firms repeats the lessons that have been stressed in Chapters 2 and 3 about how the team should be constructed and cross-fertilization encouraged. But relative to this chapter, the more important principles are how the Japanese car manufacturers provided their suppliers with incentives to learn and to innovate, whereas the American car manufacturers did the opposite. In this context, one perceives the most pernicious characteristics of the neoclassical model and its emphasis on productivity, and how it affects the relationships in the supply chain, even when the components are being made within the same company! In other words, the model of productivity discourages collaboration internally as well as externally.

As an example of product development, Honda created a management team in 1986 to develop a new Accord. The team leader was allowed to select the individual experts that he wanted from the various functional departments of marketing, product planning, styling, advanced engineering, detail engineering. Although the individuals maintained their contacts with their functional departments, they worked for the project leader who could directly give them orders. Despite the diversity of expertise, communication was high because the team made the critical decisions at the very beginning and thus resolved conflicts that might impede communication. Because different variations on the basic model were needed in different parts of the world, two subteams were also created: one in Japan for a hardtop version, and another in the United States for the coupe and station wagon versions. All the design work was completed, the model was placed into production within four years, and the new Accord immediately became the top seller in the U.S. market in 1989.

The dies for the body parts were developed at the same time that the team began the design of the car body. With these early specifications, the dies for the Accord were ready in about thirteen months in comparison to the average of twenty-five months for American car producers. The effectiveness of the communication between the teams can be measured by the amount of changes made as a share of the total die cost. In the Japanese producers, it was 10 to 20 percent, whereas in the American producers, it was 30 to 50 percent.[18]

In contrast, GM established a project team to design what they called the Model GM-10, which was to be shared across the four major car divisions of Chevrolet, Pontiac, Oldsmobile, and Buick. Mr. Dorn was designated as the team leader, but he could not really give orders or select the people that he wanted. Instead he became mainly a coordinator. His first problem was obtaining agreement from each of the major divisions about the basic characteristics of the platform that would be shared among the four divisions. Only tacit approval was obtained and then only after a considerable time period. GM was a classic case of silos and how detrimental they can be for product development. Later, top management would intervene to make changes, which delayed the project even more. In other words, another blockage in the development process was the centralized decision-making powers intervening at the wrong moment in the process. As explained in the report, the loyalties of the team members still remained with their original departments, not the development project, whereas in the Japanese companies, doing well in the team was not only rewarded but seen as a vital step in one's career.[19]

Thus, in summary, the Honda team was a diverse product development team with high levels of communication, whereas the GM team consisted of individuals not really working together. Instead of the team being granted decision-making powers as they were at Honda, the GM team had to engage in negotiations with the various separate powers that constituted the divisional structure of the company. The real differences, however, were in the way in which the various components of the car were designed and placed into production.

Another example of how to make the supply chain a learning and an innovation chain can be found in another Japanese car manufacturer. In the description of learning that follows, the Japanese model is based on research on the Toyota system because it was the first major company to design its supply chain as a learning chain. And it did so for historical reasons, reflecting a major organizational change. While typically the discussion in the American management literature has been on the just-in-time delivery system, the real strengths of this system were its ability to create collective learning and innovative products. Not unexpectedly, given the constant focus on productivity for determining CEO salaries, this more important aspect has been lost to view. The paradox is that Toyota, by relying upon largely separate stand-alone companies, achieved collective learning and innovation, whereas GM, which produced 70 percent of its parts internally, could not. Continuing with this paradox, one of the arguments made in the Introduction is that the idea of an innovation network—*when properly connected*—can deliver innovations faster, with higher quality, and for lower cost. Evidence for these results comes from a detailed comparative study of the speed and cost of development of new automobile models in Japan and the United States.[20] Each new model in the United States during the 1980s consumed about 3.1 million hours of engineering time and 60 months of development time in comparison to 1.7 million hours and 46 months for the Japanese automobile companies. Since then the American automobile industry has steadily reduced its development costs and time, but the Japanese have kept ahead of them. The first Toyota Prius was developed and placed into production in only 15 months![21]

Eiji Toyoda and Taiichi Ohno were the transformational leadership team that created the supply chain that gradually made Toyota the most successful automobile company in the world.[22] What are the principles that they established? The key was to treat the suppliers as partners in terms both of profit and of product development. Prices in the contracts between the suppliers and Toyota were set realistically rather than extremely low as in the American case. Furthermore, the advantages of cost reductions across time were agreed upon in advance, but if the supplier learned faster how to improve quality and productivity, then these fruits were for the benefit of the supplier. Thus, the suppliers were encouraged to learn.

When a new car was developed, the suppliers were treated as part of the development team. The first-tier suppliers assigned designers to the development team shortly after the process began. They were given specifications and asked to produce a prototype. They were encouraged at the same time to talk among themselves rather than compete against each other. They were given complete responsibility for a component of the car model with an agreed-upon performance. Working with both resident design engineers of the parent company and the second-tier suppliers, they performed the detailed development and engineering. Undergirding this system was a complex pattern of joint ownership. Toyota had equity in each of its major suppliers, and they had cross-holders in each other. Toyota also provided both capital in the form of loans and human capital in the form of transfers of managers. In other words, Toyota built a community of trust with shared responsibility and concern that each member of the supply chain make a profit.

In America, there were two models for the organization of the supply chain at the time of the study: the GM model, where 70 percent of the parts were produced internally, and the Ford model, where only 50 percent were. However, regardless of which model, in fact most of the 10,000 parts were designed in-house by the design engineers of either GM or Ford rather than in cooperation with the suppliers that built them, let alone completely designed by them as in the Japanese case. Thus, the manufacturing expertise of the supplier never became an element in the design of the part. External suppliers were asked to make bids with a guarantee of a maximum number of defects per 1,000 parts. The process of competitive bids meant that suppliers had to bid low and hope that over time they might make a profit. Nor could they make recommendations about improvements in the manufacturing of even their own parts, so that they would articulate better with the other parts in the car because the information about the specification of other parts in the car was kept secret. This system was designed to create mistrust.

In the case of GM, although many of the major components were produced internally (Harrison Radiator, Sagamore Steering, and AC Spark Plug), they operated as profit centers and developed their own internal cultures.

Separate silos developed around each of the major sectors involved in the design of a new model as well as the major car divisions that shared the basic platform. Thus, despite the common ownership by GM, the company failed to achieve cooperation between its constituent parts, that is, both the major car divisions and the major parts suppliers.

The comparison of the Japanese and American managerial practices highlights the importance of establishing a vision that emphasizes innovation and cooperation within the supply chain rather than productivity and competition. In a series of studies of Japanese firms, the extensive use of supplier networks and research networks were very important to the product development process because they allowed for continued learning of new technical skills in the supplier firms. Not only were products developed more quickly, but they also attained higher quality. Cross-fertilization of ideas was created by diverse functional teams.[23] The Japanese case also illustrates the advantages of increased focus in specialization of component suppliers. The paradox is that by emphasizing cooperation rather than profit making of the parent company, the Japanese automobile industry has achieved a much higher level of both productivity and innovation than the American car companies who pushed the productivity model to its *il*logical limits. Of course, this does not mean that there is an absence of competition, but competition has shifted to a higher level, namely, between Toyota and Honda and between them and the American automobile companies, which increasingly are probably best called the "little three."

For managers concerned about how to construct learning and innovation supply chains beyond changing the general orientations, I turned to a recent study of the Spanish automobile supply chain that focuses on some of the specific managerial practices that reduce the time of product development and also its cost.[24] Let me start with *cooperation* and how to build it. Previously, I emphasized the importance of profit sharing and incentives for learning. But important as these are, there is more to cooperation than this. In the Spanish study, cooperation could exist in the following activities:

1. Training
2. Product development
3. Process development
4. Quality improvements and reduction of environmental risks
5. Benchmarking

6. Technology transfer

7. Marketing

Not unexpectedly, the companies that cooperated in one activity tended to cooperate in all of them. Although this study of Spanish suppliers only provides a snapshot at one time point, it is consistent with the wonders of cooperation. *Cooperation tends to produce a positive evolution toward a stronger and stronger relationship or coupling, while competition has the opposite consequence.*

In the Japanese supply chain, cooperation in product development was emphasized and cooperation in process development was implied, which focuses on how to improve the manufacturing process itself. Likewise, cooperation in improving the quality of the product and reducing risk to the environment is of equal importance and closely related to the former kind of cooperation.

Most interesting from a managerial perspective is the cooperation in training and technology transfer. These are powerful practices to ensure that everyone "speaks the same language." At the network level, these practices parallel those discussed in Chapter 3.

Finally, another pair of cooperative activities is worth highlighting: benchmarking and marketing. Benchmarking means scanning the competitor's products and then setting clear performance goals so that the new product being developed is superior in one or more ways, a topic discussed in more detail in Chapter 7. The clarity of the objectives allows managers to monitor their research teams' technical progress and for marketing campaigns to sell the product.

The greater the cooperation in the Spanish supply chain, the shorter the product development time and the lower the cost. Within this general finding, the issue is which managerial practices appear to be strongly associated with high cooperation and therefore reflect what managers should emphasize. One group of practices involved multifunctional teams, benchmarking, and rapid prototyping, and another group of practices involved design for manufacturing and concurrent engineering. Not only were these practices more typical of high-cooperation suppliers, but they increased the impact of cooperation on the rate of product development. In other words, there is a multiplier effect on product development time and cost. These practices become especially critical when there is a dramatic change in the environment, such as existed in the United States relative to the development of hybrid cars.

THIRD DEGREE OF CONNECTION

Creating Research Consortia to Turn an Industry into a Technological Leader

The previous examples indicate how the principle of cooperation within the supply chain for complex products is critical for commercial success when there is continual technological evolution. Therefore, by continually learning and innovating, the supply chain allows for the industry as a whole to survive. The following success story illustrates how powerful the creation of a research consortia in a high-tech industry can be for making a sector a world leader. The research consortium SEMATECH, which combined most of the semiconductor companies in the United States, provides the specific example.

The SEMATECH research consortium originated in the 1980s when the U.S. government relaxed its antitrust laws so that business firms could form research consortia with their competitors. The federal government changed its industrial policy to correct the decline of this key industry because the Department of Defense (DOD) felt it was vital for national security.

This case study illustrates several important themes in this book. Besides the obvious one of cooperation between competitors in an industry, another major theme is the importance of a public and private alliance for restoring the innovative edge. In this instance, the major form of cooperation was that each consortium member contributed to the one-half share of the development costs that the private firms paid.

The most striking demonstration of the advantages of cooperation between competitors is the creation of SEMATECH itself, which united an entire industry in the attempt to restore the American semiconductor industry. Not only do firms need to adapt to knowledge growth, but so do entire industries. How collective learning was accomplished by this research consortium provides business managers and policy makers with a number of lessons on how to reestablish the automobile industry. The success of SEMATECH has to be appreciated by comparison with failures in other research consortia in the United States where competitiveness has prevented cooperation for the common good. One of the more notable failures was the U.S. Council for Automotive Research as well as the predecessor for SEMATECH, Microelectronics Computer Cooperation. The source of interorganizational conflict in research consortia is usually about proprietary standards. The semiconductor industry has known long, drawn-out battles in court, with those between Advanced

Micro Devices and Intel being the best known. Analyses have shown that in industries with rapidly changing technologies, proprietary standards create an intense level of competition fueled by the law of increasing returns, or first mover effects, that is, the one with the first technological advance captures more of the market.[25] But we have also seen in the case of IBM, and later in that of Ericsson, how under certain conditions outsourcing becomes a major way of stimulating the use of a product.

So, what is novel and important about SEMATECH for policy makers is how much it in many ways parallels the lessons discussed in previous chapters about the characteristics of a new model of organization; but in this instance, the lessons concern the characteristics of a new model for research consortia:

- Complex charter and multiple sources of resources
- Transformational team of leaders and a separate interorganizational unit
- Practices for stimulating cross-fertilization and cooperation

These characteristics contributed to the huge success in restoring the innovative edge in the semiconductor industry, although several additional explanations can be provided as to why America regained the lead in this industry.[26] The key insights are the following:

1. SEMATECH offers a demonstration of how the supply chain, or vertical relationships, can be changed from one of exploitation and conflict to one of cooperation with benefits for all. One of the most important accomplishments of the consortium was improving the manufacturing of the equipment needed to produce chips and establishing cooperative relationships between the suppliers and the chip manufacturers. Some argue that this focus helped to reduce the competition between the chip manufacturers because each of them benefited from improvements in the supply chain of equipment.[27]

2. What makes the charter complex is that the initial strategy of SEMA-TECH was to build the infrastructure by (a) improving the supply base of equipment and materials, (b) improving manufacturing processes, and (c) improving the management of factories. The main objective was to restore American competitiveness in this high-tech sector. In one sense, in this industry, the benchmarking of what is needed was relatively simple because the pathway of technological evolution was so apparent. The industry had to increase the number of viable chips that could be obtained from each six-inch wafer, and

thus emphasize miniaturization, as well as increase the capacity, or power, of each separate chip. The latter is accomplished by reducing the width of the circuit lines on each chip. Similarly, these desiderata apply to a number of other industry sectors, that is, make the product smaller and give it more functions. In fact, the capacity to design entire systems on single chips is transforming many other industries.

3. Given the themes of this book, the most interesting aspect of the consortium's objectives was recognizing that the narrow productivity goal of lowest cost was not viable. So, instead, the emphasis was placed on the total-cost-of-ownership criterion that took in consideration the quality costs discussed in Chapter 1, such as costs of installation, servicing, and reliability during manufacturing. In other words, they focused on quality in the sense of reducing operating costs and made that an important set of research objectives. Even more critically, everyone realized that the standards had to continually change. In other words, they adopted an evolutionary model of constant improvement of the standard, something new within the context of American manufacturing.

The SEMATECH consortium began in 1987 with fourteen members that together accounted for about 80 percent of the American production. These fourteen members contributed $100 million a year and the federal government matched it with the same amount. The formula for each participating organization was 1 percent of sales. The government paid its share from the Defense Advanced Research Projects Agency (DARPA), reflecting the concerns of the DOD about not having an independent semiconductor industry. Thus the consortium had available a budget of $1 billion over the five-year duration of the consortium, most of which was spent. This diversity of resources meant that the consortium was not dependent upon any one private or public source, which gave it flexibility. What is perhaps most interesting is that this budget reflected only about 5 percent of the total R&D budget for the industry, indicating how much of a multiplier effect consortiums can have when properly managed.

It would be a mistake to think only in terms of financial resources. Some of the participating firms' top executives donated considerable amounts of their valuable time to convince members of the industry to participate. Furthermore, the firms also contributed their own research personnel and thus sets of expertise. Another key resource that was contributed by the various members was what might be called "best practices" relative to various managerial issues, including conflict resolution, communication, decision making, and the like. SEMATECH is an example of how a research consortium represents a knowledge pool from which, relative to a particular situation, the best practice could emerge and all would learn from it.

What success did SEMATECH have in its first five years? Relative to the consortium goals and as measures of its technical progress, the width of the circuit line was reduced from .80 microns by March 1989, to .50 in November of the same year, and finally to .35 in December 1992, increasing the power of each chip accordingly.[28] By 1992, at the end of the first five years, American manufacturers had 44 percent of the global market in comparison to Japan's 43 percent, a considerable improvement over the dire predictions that they would only have 20 percent.[29] Because of this success plus the efforts of the then-current CEO, Bill Spencer, eleven of the original fourteen members of the consortium agreed to continue for another five years with the same financial arrangements as were made for the first five-year period.

The visionary leadership team of three executives was called the Office of the Chief Executive. They conferred frequently, attended important meetings together, and generally worked closely together.[30] This pattern was duplicated throughout the interorganizational unit, as is indicated later. The importance of differences in leadership is illustrated by the relative strengths and weaknesses of several members of the leadership team. Bob Noyce was a charismatic figure who came out of retirement. The first day he arrived at SEMATECH, he took off his tie and established an informal dress and operational style. But it was difficult for him to narrow the focus of the various technical objectives created in the long list established by the work groups at the beginning of SEMATECH because he was broadly supportive of what everyone wanted to do. In contrast, Turner Hasty was the member of the leadership team that tried to keep the focus on a limited list of objectives.[31] He was also subtle in how he obtained information from individuals and lowered the sense of secrecy that prevailed in the beginning. He believed that the engineers would not provide him with the technical information needed to make decisions in the beginning before much cooperation had become established, so he presented a plan that was not specific enough. The engineers criticized the plan and, realizing at the same time that they all knew basically the same thing, gave Hasty the information that he needed.

After an initial start in 1987, the interorganizational unit of coordination was established in 1988 in Austin, Texas, in a four-story building with a small fabrication plant. About two hundred employees from the various companies were contributed to the workforce, usually for a period of several years. In addition, permanent employees were hired, bringing the total to about six hundred employees in the first few years and toward the end of the first five-year period, eight hundred employees.

Besides the key elements of a leadership team that was transformational and the separate interorganizational unit in which the research was housed, several other key features of its governance should be stressed because they can provide models for other consortia. The first is the emphasis on joint problem-solving in meetings at all levels and in the various committees and councils in the complex governance structure that defined SEMATECH.[32] The second is the important elaboration of technical advisory boards. Usually, discussions about government in the management literature focus on the administrative side, such as the policy board and the CEO, and ignore the technical side, which is much more important for attempting to regain technical leadership. Therefore, the consortium provides a good example of both these principles of organization.

The joint problem-solving started even before the consortium became operational. Some thirty workshops with company representatives developed a long wish list of technical objectives, or technological roadmap, that included such topics as manufacturing process development, lithography, front-end processes (doping and thermal processes), back-end processes (etch and film deposition), packaging, and so on.[33] These workshops were, of course, an important mechanism not only for setting specific technical objectives but obtaining "a buy-in" from each of the companies. Everyone had their favorite technical objective on this considerable list.

In addition to the policy board composed of members of the participating firms, an executive technical advisory board set RDT (basic research, applied research, and product development) priorities. Several other technical boards advised on specific research projects within the general priorities. Task forces and councils focused on highly specific problems, including quality, supplier relations, and technological transfer. The membership of all these boards came from the participating companies. Facilitating joint problem-solving was the consortium structure that consisted of only three levels: the directors, the managers, and the project managers. The key component was the research project. In other words, this organization was decentralized in both its structure and operation.

One important element worth mentioning, the foundation of the consortium, is that a number of the key actors had worked together at Fairchild and had developed personal bonds and considerable trust. This is not to say that there was not at times conflict in the leadership team. In the first six months, one of the directors, Paul Castrucci, made a number of private decisions without consulting everyone and was forced to leave. However, this conflict helped establish the principle of joint decision-making and transparency, hallmarks of the way in which decisions from then on were made.

Creating an atmosphere of joint problem-solving, maintaining a relatively flat organization of three levels, and selecting a transformational leadership team are all practices that encourage cross-fertilization and cooperation between companies that are competitors, as we have seen in the previous chapters. But these were not the only practices that stimulated cross-fertilization and cooperation in SEMATECH. Four other practices are worth particular emphasis:

- Having a mission of survival or a common problem united by a global vision
- Creating a dictionary of terms or a common language
- Eliminating status barriers or maintaining common working conditions
- Using conflict-resolution techniques

At first, many of the participants who were competitors and had been raised within the American competitive context with its ethos, found it difficult to work together in the same consortium. But all of them were losing market share and at a fast rate. So it became a question of survival. This mission of survival was continuously reinforced through such statements as "If it is not competitive, change it."[34] The crisis thus created a vision that transcended the specific circumstances of each company. The issue was the following: Would the industry survive? This shift in focus from the individual company and its survival to whether or not the industry would survive led to cooperative behavior and provided the larger vision that I have emphasized in stimulating cross-fertilization. Indeed, this is why it is important for business leaders and policy makers today to realize that American high-tech industry is in a crisis and that only cooperation will help the country's standard of living to be maintained.

But other aspects of the vision also facilitated cooperation and dampened the problem of competitiveness. Starting in 1990, Noyce shifted the emphasis of the consortium more and more to strengthening the equipment manufacturers. Because of concerns about the technical information of each equipment supplier being shared across their competitors, the consortium developed the technique of helping each equipment manufacturer without releasing any of its particular technical information, thus being an "honest broker."[35] This shift in focus allowed for both horizontal and vertical collaboration in the industry, and at that moment the consortium truly had an industry-wide focus. Also this meant giving much more attention to manufacturing and quality research consistent with the idea of reducing total operating costs.

Despite its being a single industry, the cultures of the participating companies varied considerably in their relative strength. More critically, their technical language varied as well. To create a common vocabulary, a dictionary was constructed in 1988 and then updated some six times in four years.

Like the Pasteur Institute, there was no formal organizational chart, and from time to time one was invented for external purposes. Some of the researchers who came from cultures with strong status differences at times found the lack of titles to be difficult. So Noyce suggested that they invent their own. All offices were open and the partitions between them only went up partway. In addition, probably the biggest example of status, reserved parking spaces did not exist.

Perhaps one of the hardest ideas to accept is that in research a clash of ideas inevitably occurs. SEMATECH was no exception. One of the major practical sources of conflict was Hasty's attempt to narrow the wish list produced in all the working groups—which was in part the initial reason for cooperation—down to a manageable size. Meetings over this process produced conflict. As a consequence, conflict-resolution techniques were used. In the use of these techniques the richness of experiences and organizational practices involved in a consortium comes into play. In this instance, Intel had developed the technique of constructive criticism and taught it to the other organizations. The key was to focus on the ideas, not the person.

The evolution of the semiconductor industry over the 1990s provides still another test of how the growth in knowledge changes the nature of the idea innovation network. The growth led to focusing increasingly on the making of a large number of different kinds of chips, reflecting again the general tendencies toward customization. A number of small companies that only designed chips, called "fabless" firms (i.e., they did not fabricate, or manufacture, the chips themselves), emerged in the United States with this technical expertise.[36] The design problem was to place a separate system on a single chip, which is how customization manifests itself in the semiconductor industry.

The emergence of these new kinds of research organizations in the semi-conductor idea innovation network was facilitated by the abundance of venture capital. So, the growing capabilities to design different kinds of chips met a differentiation of needs in the marketplace. The emphasis on the varieties of chips led to the increasing importance of the manufacturing of boutique chips. For the United States, this growth in fabless firms has meant a considerable growth in this specific research arena. In 1990, there were only seven firms with $1.6 billion in sales, but by 1997, the number of companies had grown to forty and the value of sales was $6.3 billion.[37]

The growth in knowledge also meant an increasing globalization of the industry and fundamental changes in the membership of SEMATECH. One consequence was that the next radical technological step became more and more difficult and costly. Intel had by itself developed the 150 mm wafer, and IBM the 200 mm wafer. But the attendant costs of the 300 mm wafer forced both national and international chip manufacturers to collaborate. The very high cost and risk (wrong technical solution) attached to the development of this next generation of wafers on which many chips are made have led SEMA-TECH to be internationalized.

The steps toward globalization began with the end of the second five-year period of support by the federal government.[38] In 1997, a subsidiary called International SEMATECH was founded to include five foreign firms, one European and four Asian (but not Japanese firms because of political concerns). The specific objective of this subsidiary was to develop the new 300 mm wafer chip. Then in 1999, this subsidiary was dissolved and incorporated into the original consortium, which means it now has nine American members and five foreign members. Foreign members would have access to all technical programs, including interconnect, front-end processes, assembly and packaging, design systems, and manufacturing methods. At the same time, a broad agreement was struck with Hitachi, thus making the consortium truly international. Meanwhile, the decline in American members reflected in part the feeling that the consortium no longer met some of the specific goals of its participants and the increasing concentration of the manufacturing of chips in three American companies: Intel, Advanced Micro Devices, and IBM.

But despite the seeming success of SEMATECH to save a good share of the American semiconductor industry, we again have an example of how Asian countries have been moving up the supply chain toward the most profitable components, including design. The U.S. share of global semiconducutor production has fallen from 42 percent in 1980 to 30 percent in 1990, and to 16 percent in 2007.[39] For the latest generation of wafers, 300 mm, the U.S. share declined to 20 percent in 2004. Although U.S. firms still account for 48 percent of global sales and remain the undisputed technological leaders, the U.S. share of the global manufacturing of semiconductors continues to decline. Again, as with Apple, we have a "success" story but one that is misleading if it does not focus on the evolution of market share and the loss of employment. SEMATECH's story is another argument in favor of a U.S. industrial manufacturing policy.

In summary, it is possible to save an entire American industry with a research consortium, but it does require both private and public cooperation, transformational leadership, a number of practices for encouraging cooperation, and a strategic vision that considers the entire industry. In this case, the emphasis on manufacturing research and quality research with the equipment manufacturers made cooperation between the manufacturers easier. At the same time, it illustrates the relentless pace of change caused by the evolution in knowledge and how high costs have resulted in the movement toward a global consortium and more and more production being moved offshore. Despite this evolution, this scenario offers a pattern for other high-tech industries that have only a few firms. The implications are briefly considered in the last section, in which we suggest that the United States might find it profitable to participate in other sector consortia in arenas in which it has lost some of its technological edge.

FOURTH DEGREE OF CONNECTION

Turning Regions into Learning and Innovation Networks

As we have moved from one kind of connection to another, we have increasingly focused on more and more complex arrangements, starting with various kinds of joint ventures or alliances that only involve one or two research arenas, to the connection in the supply chain, and then to connections in the entire idea innovation network attached to the components of the supply chain. In this fourth kind of connection, the focus is on creating regional spaces for learning and innovation, that is, spaces that can transcend a single industrial sector. Increasingly, companies like to locate in regional spaces where colocation exists with other major research centers and both the idea innovation network and much of the supply chain are located.[40] This becomes still another remedy for overcoming the obstacle of gaps in the idea innovation network, especially between sectors, and restoring the innovative edge.

Perhaps the most attractive model for creating strong links in all the parts of the idea innovation network is to create a Silicon Valley, where all the arenas of research are colocated along with both national and international firms and where there is not just one sector but adjacent sectors as well. Certainly the success of America's Silicon Valley has stimulated many imitators, but few of these have actually been successful.

What are the special conditions that can make a region a Silicon Valley? I would argue for a minimum of three necessary preconditions:

- Patterns of collaboration that encourage the formation of small high-tech companies
- Concentrations of diverse human capital skills appropriate for these companies
- Willingness to take high risks on new technological products and service

The most important is the first: some company that starts a pattern of collaboration in encouraging technological innovation and, in particular, encourages the formation of new small high-tech companies. The key in the case of Silicon Valley was the instrumental role played by HP that helped bring together venture capitalists with entrepreneurs who had a bright idea. Besides the heavy investment of the U.S. government in developing computers and the Internet, much of the vitality of Silicon Valley emerged because each firm after HP continued to make it easy for individuals with good ideas to start their own small high-tech firms. The firms would help those with ideas find investors, and plenty of venture capital was available. Failure did not prevent someone from starting a second company or becoming a serial entrepreneur. This cooperation in helping others start businesses naturally encouraged a high level of trust between firms and also a high level of risk taking, as in fact the consequences of failures were considerably reduced. Stanford University supplied the diversity of human capital skills that allowed Silicon Valley to take off and probably contributed to the creation of trust because so many of the graduates from their electrical engineering department launched their careers in Silicon Valley.

What prevents a "Silicon Valley–type" collaboration from developing is again the competitive model that is less interested in the collective good and more in individual firm profit. The failure of Route 128 next to MIT is an example of how important it is for one company to establish a pattern of collaboration. Without this, the two other factors do not allow for the development of a rich learning space that evolves with time, as Silicon Valley did with the dot-com companies, only the more recent wave of small high-tech companies to originate there. The example of Route 128 also demonstrates that the close presence of a high-tech research university may be a necessary, but not a sufficient, condition for the creation of a region where cross-sector learning and innovation can occur.

The best proof of these assertions is to look for other cases where comparable policies have helped to establish a Silicon Valley–like region of learning and of innovation. Hence, in this section, the case of how Ericsson built a Silicon Valley in Stockholm holds special interest for policy makers in the various state governments of the United States concerned with replicating the success of Silicon Valley.

Ericsson is another example of a company that previously enjoyed a quasi-monopoly position, providing telecommunications equipment to the Nordic countries.[41] In the late 1990s, as data communication networks began to merge with traditional switching equipment, the firm had to develop a new systems-integration language, which it did, calling it Erlang. At first, Ericsson kept this language proprietary. But when the management realized that the language had the potentiality of solving problems in multiple industries, Ericsson decided to make it an open source. Of course, one advantage was that when third parties improved the language, Ericsson benefited as well. In other words, Ericsson abandoned its propriety rights for the same reason that IBM did: It realized that open sourcing facilitated collaborations and was worth the price.

But the more striking example of a strategic change is how Ericsson established a venture capital fund for start-ups and at the same time changed its personnel policies. Previously if its engineers left to work at start-ups, they could not return to the company. Now Ericsson announced that it would re-hire individuals when the company they worked at failed. With an industry-specific language and the encouragement of job mobility between companies, the engineers were more willing to work for the new companies (i.e., the risk had been reduced). The consequence was the creation of hundreds of companies, most of which were working on specialized software for the Internet. Attracted by this diverse pool of human capital, Nokia established a research center in Stockholm while Microsoft opened an R&D software center to concentrate on the Internet. In other words, through its policies, Ericsson was able to create another Silicon Valley in Stockholm by overcoming the path dependency of avoidance of risk taking by its engineers. And Ericsson has profited from these changes: It now dominates the end-to-end wireless communication systems with about 40 percent of the market for third-generation wireless equipment. Although Sweden was not part of the study reported in Chapter 1, it substantiates, for telecommunications in this country, the evolution of the idea innovation network and how to overcome failures in evolution.

However, the example of Ericsson reflects the decisions of a business firm. Silicon Valleys can also be created by decisions of the government with this objective in mind. Taiwan built its own highly successful Silicon Valley with a strength in the design and manufacturing of chips.[42] This was done by establishing an industrial park in which a research center, Technology Research Institute (ITRI), was constructed. Close to it a technical university was established. Small firms were encouraged to locate in the industrial park. Finally, the government wisely decided that the future for the country was to focus on the design and manufacturing of what are called boutique chips, that is, chips designed to meet certain functions, or the customization strategy. ITRI has research collaborations with companies, universities, and governments all over the world. It has followed exactly the model of starting with the simplest part of the supply chain and working its way to the high-value-added components, from test and assembly to wafer fabrication to design.[43]

These policies have made Taiwan a key player in the internationalization of SEMATECH (described previously), and have propelled Acer into its position as one of the leading manufacturers of laptops as noted in the Introduction.

Another way of creating a Silicon Valley is illustrated by Finland's decisions to develop rapidly a number of technical training sectors at both the secondary school technical level and the university engineer level as the telecommunications system evolved. It created Otaniems Science Park with a technical university and major research centers.[44] As a consequence, both Siemens and Ericsson located major research centers there. Of course, one powerful attraction was the enormous success of Nokia, another organization that has built a global corporation by continously changing the standards for cell phones. Until Apple began to replace it technologically, it had 35 percent of the global market.[45] All together, some ninety technology and service companies have located there, and they employ eight thousand professional researchers. The result is a rich intellectual space that has not only maintained Nokia's commanding position but made the region a world leader in microelectronic mechanical systems (MENS), that is, valves, gears, mirrors, and parts of semiconductor chips.

In summary, Silicon Valleys can be encouraged by following the practices indicated in these various examples. Key practices are focusing on creating a climate of trust, encouraging risk taking, and concentrating on a diversity of specialties and disciplines.

BUSINESS AND POLICY LEARNING STRATEGIES FOR A TYPOLOGY OF SECTORS

The thesis of this book—that each economic and noneconomic sector is dissimilar—implies distinctive alliances of private firms and public governments with different strategies for restoring the innovative edge in each sector. Although the evidence supports this thesis, there are some family resemblances, provided one has the correct dimensions for creating a typology of economic and noneconomic sectors. The major economic dimension for describing industrial structure is the number of firms, with some sectors dominated by a few large firms, or oligopolies, and those sectors with many small firms, where market competition is strong. The major innovation dimension for describing industrial structure is between those sectors that are high-tech with large investments in RDT, 4 or 5 percent of gross sales, and those that are low-tech (with investments of less than 4 percent). In the former case, the idea innovation network has differentiated in various ways, and in the low-tech sectors evolution has not as yet proceeded that much. The concept of the innovation network allows us to considerably expand the traditional economic ideas about industrial structure. In addition, the innovation perspective adds various noneconomic sectors such as health, education, and welfare, which can also be classified into high-tech, as is the case in health, or low-tech, as in the case of welfare. But once one shifts to the noneconomic sectors, the word *firm* is no longer applicable. Instead, the focus is on the service providers and, beyond this, the delivery system.

Together these two dimensions, which themselves include a number of ideas, generate a typology of four kinds of economic and noneconomic sectors:

1. High-tech sectors dominated by a few firms or service providers, nationally or globally
2. High-tech sectors with many firms or service providers nationally
3. Low-tech sectors dominated by a few firms or service providers, nationally or globally
4. Low-tech sectors with many firms and service providers nationally or locally

Examples of the first type are aircraft, automobiles, chemicals, defense weapons, nuclear energy, pharmaceutical companies, train transportation equipment providers, and so on because I am using 4 percent of investment in R&D as the cutoff point for high-tech. Obviously, this is a dimension and would move from four types to nine, and so on. In the public sector are the mission agencies of National Aeronautical and Space Agency (NASA) and National Oceanographic and Atmospheric Administration (NOAA) and the large public research laboratories attached to the Department of Energy (DOE). The second type is represented by alternative energies other than nuclear, computer software, life sciences, advanced materials including nanotechnology, universities that include a number of different disciplines, and the like.

Some of the low-tech industries that are dominated by a few firms are food products, tobacco, paper, and primary metals. Their counterparts with much less concentration are the sectors of agriculture, apparel, wood products, furniture, printing, and so on. The noneconomic sectors in the fourth category include many of the local services that we take for granted, such as primary and secondary education, police and fire protection, and welfare services of various kinds. Within each of these categories, one can use a more fine-grained approach representing market niches that are not necessarily dominated by the same firms or even the same countries. For example, two companies, Boeing and Airbus, control the production of large civilian aircraft. But this ignores market niches where there may be more firms and not necessarily American ones. Brazilian, Canadian, Chinese, and Russian firms dominate the medium- and short-range civilian jets niche. Small executive jets are still another market niche to say nothing about helicopters and military aircraft of all kinds. Conversely, computer software has a few large companies such as Microsoft, but in fact there are many different kinds of software companies, most of which are highly specialized. One of the more interesting new niches is animation media.

The potential advantage of this typology is the notion that one sector that has been successful in keeping ahead of the curve can become a model for those other sectors within the same general type that are experiencing difficulties. Usually, however, when models are adopted, not all of their features are necessarily copied. So in the following discussion, some attention is paid to what I consider to be the salient features. The ideas expressed are somewhat speculative, but they are intended to start a national conversation about how best to restore the innovative edge in a number of our economic and noneconomic sectors.

Given the recent bailout, probably the sector that most concerns the American public and policy makers is that of automobiles. The model for this sector would be SEMATECH. As already discussed, the key features in this model are a transformational leadership team and a very complex decision-making structure that would focus on what are the most important scientific and technical problems to solve for the cars of the future. But, although a consortium was assembled for this sector to develop a new car, it was not organized along these principles and the automobile industry did not really provide the kind of expertise that was needed. Again, the reason is probably that this sector has been most dominated by the ethos of competition and measures of productivity rather than innovation during the past forty years.

The comparison between the Japanese supply chain and the supply chain of GM indicates one component in which drastic changes have to occur in firm behavior. The key is to make the *supply chain* a learning and innovation chain. In particular, the automotive industry is one where the idea of a third industrial divide has perhaps its greatest promise given the varied performance characteristics of automobiles that are desired by the public. And although the idea innovation network was not emphasized in my discussion of the automobile industry in the Introduction and in this chapter, it is highly relevant in view of the many scientific and technological problems that have to be solved, not only for the creation of a third industrial divide but also for developing the different technologies required for the disparate kinds of vehicles that should be made available for ground transportation. The hydrogen car represents probably the most extreme example. But it is not the only one. Another example is the development of a grass that has a high yield of ethanol fuel that can be grown in the desert, for example, to reduce dependency upon oil, which will not impact on food prices too much by using scarce agricultural land. A number of interesting experiments are being undertaken. But if these many efforts are to pay off, the research consortium should coordinate and integrate these efforts and be sure that there is research in the six arenas because distribution or commercialization present problems.

Besides the automobile, another major sector of ground transportation that has recently moved back on the American agenda is high-speed trains. The United States is so far behind in this technological sector that perhaps it will have to import all of the equipment, rails, and expertise. At the same time, it might be possible to enter into a research consortium with a major supplier, whether French, German, or Japanese, that would develop the next generation of high-speed trains, ones that consume much less energy and are much cheaper to construct. Given the size of the North American market, this may be a distinct possibility. One critical issue is how to construct these systems to exploit existing railroad tracks (e.g., built stacked above existing tracks to allow freight trains to pass underneath) so that this vital part of the national transportation system is not disrupted. Again, these problems provide opportunities for getting ahead of the innovation curve. What has been said about high-speed trains can also be applied to another niche in ground transportation, metro and tram systems.

Little has been said about nuclear energy as an alternative to oil for generating electricity. With global warming, this has again become an alternative form of energy that clamors for more research, despite resistance from some ecology-minded groups. An international consortium with the French, who have had the longest and largest experience with the maintenance and operation of nuclear energy plants, to create the next generation of nuclear power would seem to be a distinct possibility. The French have made a number of modifications to the Westinghouse patents, which they eventually bought. It could also be an area for U.S.-Russian cooperation. Furthermore, the problems of waste disposal and safety require global solutions, and the cooperation of any countries that use this technology will be needed. If some of these problems can be solved, the potential market in this area is also huge.

The defense industry has long been one of those sectors where the idea of an international consortium à la SEMATECH seems most bizarre. But again, if we return to an argument in Chapter 1, that war has to be customized and that different forms of wars are fought with disparate technologies, then there are opportunities for the development of a research consortium attached to the North Atlantic Treaty Organization (NATO) that would concentrate on two kinds of war technologies: civil wars and genocide. In both instances, the technological problem is how NATO can maintain peace and prevent genocide. The equipment needed varies by terrain but inevitably requires new radical technologies to reduce the loss of life, if we take the case of Darfur as an example. In this area, the United States has acquired a great deal of expertise in the wars of Iraq and Afghanistan that it can share. Because a number of NATO countries have been involved in these wars as well, to say nothing about the civil war in the former Yugoslavia, the pooling of this experience could be especially fruitful.

These four examples do not represent all of the high-tech sectors that are dominated by a few firms, but they illustrate how the model of the research consortium, and in particular an international one, might help restore American expertise in these sectors. Clearly, the United States has much to contribute even in those sectors where it is technologically behind, given the extensiveness of our national research laboratories and the very high investments in RDT. In addition, our market is large and attractive. But the key is for these consortia to work on the products of the future that advance technological sophistication, improve quality, and reduce energy consumption and are customized to the particular demands or circumstances of various countries. At the same time, however, while the general model of SEMATECH has been suggested, it is important to recognize that each of these requires different specific strategies and partnerships, not only between the public and private sectors but their counterparts in other countries.

When we shift to the large research organizations that are dominant in the high-tech noneconomic sectors, a different set of problems emerges. In this context, the problem is lack of connection to the private industry that could effectively exploit the many scientific breakthroughs and technological advances occurring in the many national laboratories connected to the DOE. One of the major functions of these laboratories is to build platforms for research by industry. And although industry takes advantage of these opportunities, the question still remains whether more radical innovative products and processes might be developed if there was more cooperation between the research teams in the public research laboratories and the research teams in private industry—an issue that I return to in Chapter 6 on industrial policy. Although each of the DOE laboratories has various kinds of outreach programs, I would hypothesize that the connections between these and private industry are not as strong as they could be. Recently, the DOE created four nanotechnology research centers, which might be a good place to experiment in how to accelerate the development of new technologies via public-private cooperation.

Many of these sectors—alternative energies other than nuclear, specialized computer software, life sciences, advanced materials including nanotechnology, disciplines found in universities—are so different that it is difficult to provide a single overarching principle about how they might be organized to achieve higher rates of radical product and process innovations. One of the more startling findings reported in the Introduction is the negative balance of trade in the life sciences. Given our huge investment in medical research, it implies that the valley of death is especially strong in this sector which was diagrammed in Chapter 1. Evaluation studies in these areas by the federal government, again a topic for Chapters 6 and 7, that would locate all the relevant actors might be most helpful in facilitating more innovation by identifying various gaps and, in particular, the obstacles and blockages that are preventing more rapid development of radical innovations.

Earlier in this chapter, we observed that one major blockage in some of the high-tech areas with growing complexity and difficulty of research problems is the lack of cooperation between universities, in part because of an institutional rule of NSF and in part because of the disparate rules and regulations of the participating universities. Because problems are becoming more complex and difficult, learning how to facilitate cooperation between research organizations doing basic research should be a top priority. In this regard, there may be lessons to be learned from the research on team science in the National Institutes of Health (NIH), which I do not address here.

With the Obama administration, alternative energies are receiving a great deal of attention because one of the national laboratories attached to DOE specializes in the basic research involved in them. But again, the ethos that national laboratories must be at "arm's length" from the private sector has diminished a great opportunity for greater cooperation between the National Laboratory for Alternative Energies and any number of small high-tech companies. This is one of the areas where an international consortium, especially with Denmark and Germany in the area of turbines, might be effective. The Chinese are aggressively moving into this area, and only a consortium might be able to protect jobs both in the United States and in Europe. And although considerable progress has been made in voltaic cells, a consortium might accelerate the process and also solve the problem of improved efficiency in thermal couplings, one of the weak points. As we noted in Chapter 1, a major problem is creating new kinds of electrical grids that can incorporate the different ways in which electricity is produced. The spread of voltaic cells and the selling of the excess electricity to the companies that provide electricity have many exciting possibilities.

By contrast, in the low-tech sectors, the neoclassical paradigm is much more powerful. Thus a policy that emphasizes getting ahead of the innovation curve is likely to have the biggest payoff. Let us start with a few of the major sectors in which a few large companies control the sector. Coal is one sector where considerable effort is being expanded to see if it can be developed as a clean fuel. One wonders if basic research on converting coal into material for the manufacturing of automobiles might provide a way of protecting the employment of miners without damage to the environment.

The wood pulp industry is another sector where the United States is far behind its European competitors in the efficiency of its processes and in protection of the environment, and it remains to be seen whether this industrial sector will disappear. Here is where an international consortium, especially with Sweden, which has a superior technology, might improve productivity and reduce risks to the environment. Admittedly, part of the reason for the failure of the United States is the differences in regulations across the various states. The major firms say they have been reluctant to build new automated facilities because of the problem of regulations.

In contrast, the furniture industry is alive and well. It is a sector with a considerable range in the size of firms, where both many small companies and a few large manufacturers are common. One of the reasons for this sector's survival is that the purchase of furniture is largely controlled by national tastes, which protects American industry to a certain extent. But given the general popularity of American styles and culture, one wonders if this industry might be better organized so that it could export certain distinctive American styles of furniture to the developed countries. In competition, Italy has saved its furniture industry with constant improvements in design and in materials. It exports worldwide and sets the standard for taste. The federal government, in cooperation with the state governments that have sizable representation in this sector, such as North Carolina's, might want to establish a research center that would further the design and quality of materials and, most critically, conduct commercialization research on the interest in American-style furniture overseas. Some candidates for potential sales overseas are the Wild West, Arts and Crafts, and Shaker furniture styles, all which are famous for their simplicity.

Another industry that we have largely lost but might be restarted is the machine tool industry. One of the reasons that we lost this industry was the overemphasis on general machine tools rather than the specialized ones for certain sectors. The emphasis on customization and the need to create new machines that are much more energy efficient could provide an opportunity to restart our expertise in this sector. To do so would also require training the specialized technicians that could operate the machines. Perhaps the most fruitful approach would be for different community colleges to focus on disparate sectors that are well represented in their state and with local, state, and federal funds develop research centers and training centers that could provide the process innovations and the individuals that could exploit them.

The construction industry also comes in all sizes and shapes. At one extreme are the local home builders and remodelers, and at the other extreme are the international companies that build huge projects such as dams, bridges, roads, and skyscrapers. The refitting of homes, churches, hospitals, schools, and various public buildings—an area where Germany is considerably ahead—so that they are much more energy efficient would create an enormous boon in our construction industry and create a large number of jobs. But more research still needs to be conducted on finding the best ways of refitting existing buildings at the lowest possible cost.

A major idea advanced in Chapter 6 for facilitating innovation in the low-tech industries is the construction of extension services in the Department of Commerce for the low-tech industries, and especially those with a multiple small companies. Again, a grand alliance between the private sector, local and state governments, and the federal government would have to determine which sectors have both the greatest need and the greatest opportunity for stimulating innovation that can gain market share nationally and globally. General themes that work across a number of sectors might facilitate the transfer of ideas from one sector to another.

The themes of the concentration of the industrial structure and the expenditures on RDT allow us to observe family resemblances across some of the sectors within one of the four types. But as the discussion of various sectors indicates, there are still considerable differences in the kinds of strategies that should be adopted. As I indicated at the beginning of this section, a number of these ideas represent speculations about what might be possible. The intent is to start a conversation about the future of the United States and how to create new strategies that will protect and create employment.

EVOLUTION, LEARNING, AND NETWORK COORDINATION

At the conclusion of Chapter 2, I mentioned the new knowledge paradigm of organizations. Since then, each of the chapters has made a contribution to this paradigm by emphasizing how organizations can learn. In Chapter 3, it meant the adoption of managerial practices to encourage cross-fertilization of research teams, and in Chapter 4, the issue is how to create communication between departments and divisions within the same organization. This chapter in particular indicates how important it is for the leadership to take an aggressive stance toward the environment to create a learning organization: building disciplines, expanding the fontiers of knowledge, and solving national and global problems. All of these strategies create opportunities to learn. Chapter 5 continues this theme but moves to the next level, learning within the sector, whether the idea innovation network or the supply chain. Cross-fertilization at this level is far more difficult and requires special arrangements, of which a number of examples have been provided.

These examples also illustrate how the thesis of open innovation can be implemented systemically. One has to eliminate the various barriers such as existed in the universities and GM. In addition, it is critical to build bridges toward differentiated arenas, as described in Chapter 2, and also with the technically sophisticated components in the supply chain.

The many examples in this chapter also illlustrate the following themes in this book:

- Obstacles and blockages explain failures in evolution and lack of response to changes.
- Networks are better than markets for exchanges of scientific and technological information needed to create radical innovations.
- Distinctive kinds of networks connect various parts of the idea innovation network.

As I indicated in Chapter 1, evolution does not occur automatically, but the model about how it unfolds provides insights about what has to be done to encourage it. The major issue is to change organizational and governmental policies and strategies. American universities failed to achieve innovation because of the obstacles of conflicting organizational policies and procedural blockages as well as NSF's policy of not funding researcher time. In contrast, both IBM and Ericsson dramatically shifted away from their previous policy of tight control over their intellectual property and went to outsourcing, which encouraged the spread of their products. Ericsson went the extra mile and encouraged its engineers to become entrepreneurs, and as a consequence turned Stockholm into an international center for Internet software.

These two businesses illustrate how to handle two distinct kinds of evolutionary failure: the lack of development of networks and the lack of development of new research organizations in a particular arena. These failures happened because of path dependency, the former typically found in the liberal market economies such as the United States and the latter in the coordinated market economies such as in most of Europe.[46] The case of IBM illustrates a company that shifted to a network solution to learn how to provide customized services to its clients. The case of Ericsson is an example of how a large firm created a number of small companies in a Swedish Silicon Valley. But more important, the two responses provide transformational business leaders with models for changing their businesses and in what ways. Although in general the United States does have a much better track record than Europe in the creation of new firms, especially research organizations close to universities, there are still sectors where large companies can facilitate the emergence of small companies and nurture new Silicon Valleys. Likewise, while Europe appears more capable of developing networks, in some areas it has problems in doing so as well.

In the Introduction, I mentioned the negative effects of using the performances of productivity and stock prices, and how this becomes translated into power for accountants and finance majors. These negative consequences are observed in GM's exploitive attitudes toward its suppliers and the emergence of silos around the various profit centers that prevented the kind of cooperation needed for effective learning and innovation. It may not be too late to save the American automobile industry if the lessons in the Japanese and Spanish automobile case studies are applied.

Another positive set of lessons for the automobile industry can be drawn from how the semiconductor industry overcame its obstacles and blockages via the formation of a research consortium. The specific ways in which competition within the consortium was reduced have been outlined. The importance of team leadership, an interorganizational unit, and a series of practices for facilitating cross-fertilization can be applied to other research consortia, particularly the one developing the future automobile.

Making connections between parts of the idea innovation network or between sectors is not only a question of eliminating blockages and correcting obstacles, it also requires taking positive steps to create learning regions where there are a concentration of trained individuals, a center of research, and policies that reduce risk by encouraging people with ideas to start new businesses. This requires public and private sector cooperation and a new industrial policy, which is the subject of Chapter 6.

Consistently through these many examples, the importance of developing learning networks to connect universities, firms, supply chains, and idea innovation networks has been underscored. The network becomes the preferred method for handling the transfer of difficult scientific and technological knowledge, recognizing problems, and solving them. A network does not imply an absence of market competition but only that the competition is moved to another level: sets of universities or firms and their alliances competing against others in the same area, supply chains competing against other supply chains, and national industry competing against its counterparts elsewhere. In the discussion of how the semiconductor industry has evolved, the conclusion is largely that gradually each of the high-tech sectors that have huge product development costs must become a global network, with perhaps two competitors possible, as has evolved already in the aircraft industry.

Contained in these examples are also several modal types of networks. The emphasis has been on the importance of the supply chain, the research consortia, and the learning region because of their obvious advantages for restoring the innovative edge. These types of networks can be combined in various ways. The speed of converting scientific breakthroughs into radical product and process innovations is largely dependent upon the strength of the connection between the various arenas in the idea innovation network. This strength depends upon the various kinds of managerial practices that have been discussed in this chapter: transformational team leadership, decentralization and high participation, interorganizational units, practices for encouraging cooperation and resolving conflicts, and so on. In both the research consortia and the learning region, all the examples highlight the importance of cooperation between the public and the private sectors as well, our next topic.

6 CREATING COOPERATION BETWEEN THE PUBLIC AND THE PRIVATE SECTORS

Step Seven in Restoring the Innovative Edge

PREVIOUSLY I FOCUSED on cooperation in the idea innovation network and the supply chain for restoring the innovative edge. In this chapter, I focus entirely on the need for public-private cooperation; and rather than just emphasize funding, the focal point is stimulating better *joint* manufacturing and quality research, especially of generic technologies that can be exploited by multiple supply chains. At the heart of this chapter is the argument for a new industrial policy that overcomes the obstacle of insufficient cooperation in the arenas of manufacturing and quality research between the public and private sectors.

Why is a U.S. industrial policy necessary, and why now? Four arguments can be made. First, while the United States has made some remarkable technological advances, it has not reaped the employment benefits from many of them, in part because of lack of cooperation. The example of Apple's iPod was given in the previous chapter to demonstrate how radical innovations do not necessarily translate into employment in this country. But an even more striking example is Amazon's Kindle, for which all the research and development (R&D) and manufacturing were done overseas.[1] Thus, except for the stock prices of Amazon, there is little employment gain for the United States.

In the discussion of Apple's iPod, I noted that the lithium rechargeable batteries and the display component had been developed in the United States, but for a variety of reasons we lost the technological edge in these components, components critical for multiple supply chains. Nor are these the only cases where we have gained little employment for major technological advances; other examples include the following:[2]

- Oxide ceramics
- Semiconductor memory devices
- Semiconductor equipment such as steppes
- Robotics
- Video cassette recorders
- Digital watches
- Interactive electronic games

The complexity of the reasons why we lost robotics were discussed in the Introduction and had to do primarily with poor implementation of this new technology by the manufacturers of the equipment as well as business managers using the devices for the purposes of control. In contrast, the lost of lithium rechargeable batteries has been a consequence of the low profit margins during the first cycle of a new technology and thus a loss of interest by the major battery companies that make disposable batteries.[3] Relative to the new flat-panel display technology, improvements in the quality of this component were aggressively pursued by the Asian countries because it was such a vital element in a number of their supply chains.[4] The stories vary from case to case and are more complicated than this single explanation, but with an industrial policy that focused on improving manufacturing and quality research in components of supply chains, perhaps some of these losses could have been prevented.

Second, because the logic of the evolutionary model emphasizes the need for faster speeds in the development of radical product and process innovations, the colocation of both the idea innovation network and the supply chain becomes important in those high-tech sectors where there is tight integration of the idea innovation network and the supply chain. Colocation is one reason why Silicon Valley–type areas are becoming so important and also why U.S. manufacturing has been investing more overseas to some of these areas that have been created in other countries. To reverse this trend requires an industrial manufacturing policy in which the federal and state governments and private sector firms cooperate to increase the technological sophistication of our manufacturing and the quality of key components that are involved in multiple supply chains as well as of the end products. As part of this new industrial policy, the federal government might consider some of the suggestions that were made about the possibility of improving our technological capabilities by cooperating with other countries in certain sectors where we have fallen behind. Additionally, U.S. policy makers need to consider rebuilding a number of competencies and areas of expertise, especially in manufacturing, to make our Silicon Valleys more attractive.

Third, another reason for an industrial policy is to highlight the problems of American competitiveness, especially in manufacturing. If such a policy existed, it might encourage more American firms to think about building plants in this country when radical innovations are produced, as has General Electric (GE). Jeffrey R. Immelt, the CEO, is an example of a transformation leader who is attempting to call attention to this problem of American competitiveness. He believes that the United States is losing its technological edge and cites many of the same reasons that were provided in the Introduction. For example, he believes that GE has exported too many jobs overseas and should attempt to help the U.S. economy remain competitive.[5] But to do so means confronting obstacles mentioned previously, to which he adds others such as CEOs focusing too much on quarterly reports and not attempting to improve the bad relationships between management and labor.

The implementation of Immelt's strategy to improve the competitiveness of the United States has involved three important actions.[6] First, GE is building a 350-employee plant in Schenectady, New York, to make high-density batteries to turn many locomotives into diesel-electric hybrids. Second, in Louisville, Kentucky, GE is adding a factory that will employ 420 workers to produce hybrid electric water heaters, ones now made in China. Both of these actions occurred after the labor unions accepted various concessions about controlling costs, including a two-tier wage system for new employees. Third, GE announced in June 2009 that it was building a $100 million, 1,100-employee research center outside of Detroit, Michigan. In various interviews, Immelt has argued that the only way to get better at exporting, a goal of the Obama administration, is to get better at manufacturing. With an industrial policy designed to improve American competitiveness with a focus on manufacturing, perhaps other American firms would replicate GE's example.

Fourth, the final reason for the need for an industrial policy is the *difficulty* of achieving sufficiently large economies of scale that offset the price advantages of cheap labor in the developing countries, most notably many of the Asian countries. As we have seen, this remains a serious problem in the success story of semiconductors. Potentially, there are three approaches that might help alleviate this constraint. One is to adopt some of the ideas suggested in Chapter 1 for emphasizing the manufacturing of customized products. This means re-visiting the whole issue of flexible manufacturing, which, as I noted in the

Introduction, was largely an American failure, but to push it much farther than it presently has been by creating the ability to manufacture many more parts on the same production line. Another is to focus on generic technologies and instruments that apply to multiple sectors, a point that will be discussed further in the first section of this chapter. Still a third potential solution is to take seriously the idea of a third industrial divide, as suggested in Chapter 1, and to attempt to develop this capacity along the lines illustrated in Modumetal. But all these solutions imply increasing the scientific quotient in a number of products, including those that are considered to be low-tech, which is largely in the public sector, and at the same time apply much better knowledge of manufacturing, which is the skill of the private sector. Hence, the need for cooperation.

A major obstacle to this cooperation is the opposition to even the idea of a national industrial policy. Paradoxically, the United States actually has one, even though I would advocate rethinking it. What are the major ingredients of American's national industrial policy? On the surface, the primary focus is in helping to provide the basic inputs for the development of science and technology. One important input is the amount of investment in RDT (basic research, applied research, and product development), in particular basic research to stimulate scientific breakthroughs and to fund projects that would not be funded in the private sector. The assumption is that private industry will exploit these scientific breakthroughs, or at least those that are relevant, through the normal market mechanisms. The other major component of U.S. policy is training scientists and engineers. In various ways the U.S. government, by funding research in the universities, provides substantial subsidies that help support higher education. But at another level, one much less appreciated, the U.S. government, via a series of decentralized efforts, is cooperating much more with industry than would appear to be the case.

In the concluding section of the Introduction, it was suggested that the United States needs a new policy model based on the ideas of evolution and failed evolution. One of the major differences between this policy model and the one just described is that it focuses on the process of how radical innovations are created and on detecting obstacles and blockages that explain failed evolution. But this new policy model requires collecting new kinds of data to supplement the excellent work of the National Science Foundation (NSF). A series of recommendations are made in the third section of this chapter. This new data would be vital in any attempt to discuss a new national manufacturing policy.

Throughout this book, I have at various times suggested that an important element for restoring the innovation edge is cooperation between the public and the private sectors. It is from these two vantage points that I want to suggest how to rethink our national industrial policy so that it has more of a process approach. The rethinking of national industrial policy should focus on three key elements:

1. Place much more emphasis on improving the competitiveness of the United States in manufacturing.
2. Develop cooperation between the public and private sector at the research team level and specifically in manufacturing and quality research.
3. Conduct new research studies in each sector to document how far the process of evolution has unfolded and the existence of various blockages and obstacles that might explain failed evolution.

As we discuss the remedy for overcoming the obstacle of the absence of a national industrial policy, we build on some common themes: the importance of cross-fertilization of ideas and the advantages of network coordination, but now between the public and the private sector. Together they represent the seventh step for restoring the innovative edge.

Rethinking the national industrial policy along these lines must inevitably confront a number of objections that form major obstacles in restoring the innovative edge. The goal of this chapter is to suggest reasons why these obstacles should be overcome.

RETHINKING THE EMPHASIS ON MANUFACTURING

The two major arguments against the idea of a national industrial policy that emphasizes manufacturing are (1) the world has a division of labor in manufacturing, and (2) now advanced industrial societies are service economies.[7] In a perfect world, as an advanced industrial society like the United States loses employment in some of the low-tech sectors such as shoes, clothing, and toys, as it did in the 1960s and 1970s, these lost jobs would be replaced by work in the new high-tech sectors such as computers, biotech, and aircraft. This in fact occurred. But now that we are losing jobs in these high-tech sectors, where are the next manufacturing sectors that will replace them? Perhaps nanotechnology and more broadly new products based on materials science, but every advanced industrial country is investing in nanotechnology, and there is no guarantee that the United States will win this competitive race. In fact, the U.S. expenditure on nanotechnology accounts for only about one-third of the total world expenditure.[8] The reason why the replacement of one lost sector by a new one unfolded so perfectly before is simply because the United States was the only country with large investments in RDT in the first three decades of the post–Second World War period and the sole inventor of these new economic sectors.[9] But this is no longer the case with the globalization of RDT expenditures; new sectors and niches within them are emerging in other countries. At various points, examples have been given for Denmark, Finland, Germany, Israel, and Taiwan, as well as others.

The second component of the argument, namely, the importance of services in today's economy and especially the high-tech services, also does not stand up to scrutiny. There are several problems with it. The first is that a good proportion of high-tech services are largely, although not completely, attached to high-tech products. The loss of the latter usually means the loss of the former. Thus losing a manufacturing base means losing a portion of (albeit not) all high-tech services. The second problem with this argument is that many high-tech services are much easier to be captured by other countries because of the Internet and the low cost of communication. Thus various technical information services for the computer have already shifted overseas where wages are lower. The third problem with this argument is that not only are various countries increasing their expenditures on R&D manufacturing but also they have national policies emphasizing the creation and attraction of high-tech services. We cannot expect our technical services necessarily to make up for the loss of manufacturing jobs.

Indeed, one of the main problems with the argument about the service economy is that many of the service jobs are low paying, and the shift of blue-collar workers from automobile companies to Wal-Mart means a loss of pay, benefits, and, perhaps even more, satisfying jobs that might require higher skill. Beyond this is the question—that no one has really been able to document—whether there are enough high-paying service jobs that can replace the lost manufacturing work. The evidence so far would appear to be no.

Perhaps the most worrisome problem about the service argument being the saving grace of the U.S. economy is that the best and most important jobs in the idea innovation network are being exported or created overseas, hollowing out our industrial R&D base. About 70 percent of the employment in industrial research is located in the large manufacturing companies, most of them high-tech firms. R&D work is not only extremely well paid, but location also affects where the discovery of scientific breakthroughs and technical advances that lead to the radical new product innovations and technical services attached to them occur. But, as I observed in the Introduction, now the growth in RDT investments *from our own companies* is much greater overseas than in this country. To use a metaphor from fishing, we are losing the whole food chain with our manufacturers building their research centers overseas.

Despite the argument about the service economy, manufacturing still accounts for a substantial proportion of our economy, 1.6 trillion of GDP, and 13 million jobs.[10] The objective for a national industrial research policy would be to not only retain these jobs but to increase the proportion of our economy involved in manufacturing. *Increasing our manufacturing is a realistic goal because some of the evolutionary forces described in Chapter 1 are creating opportunities for us to rebuild our manufacturing sector.* The nature of competition has changed in several ways. Because of the evolutionary forces described in Chapter 1, countries that can introduce the latest technological advance first gain enormously in the marketplace. Furthermore, they also have to manufacture the product in multiple variations to fit the customized demands and cultural differences, and with good design. Apple is, of course, the icon for this argument. The other way in which competition has changed is the way in which *productivity* is defined. Rather than it being simply the largest quantity at the lowest price, now companies that produce products that lower total operating costs over the life of the product have a clear advantage in the marketplace, as we saw in the discussion of SEMATECH. Furthermore, increasingly productivity should be defined to take into consideration the impact of the manufacturing of the product on the environment, the use of scarcer resources, and particularly the reduction in energy consumption, above all, in oil. All of these changes can reset the competitive clock and allow us to regain manufacturing jobs if we have the right kind of radical product innovations with technological sophistication, good design, multiple versions to appeal to different tastes, and the right kind of radical process innovations that improves quality over the long run and reduces the hidden costs to society.

Another argument for investing in manufacturing is that, increasingly, products and processes have multiple sector effects. For example, in Chapter 1 we observed the impact of fiber optics as developed by Corning Glass on a number of industrial sectors. The replacement of the analog processing system by digital has had similar, wide-ranging ripple effects. And of course, the

Internet is still creating radical transformations throughout one sector after another: postal service and letter delivery, bookstores, newspapers, film rental, and so on. It is this systemic component of radically new products and processes that argues for remaining in manufacturing. Conversely, the loss of high-tech products and processes, even just some components, has implications for a number of other sectors, both economic and noneconomic (in particular, military weapons, homeland security, and even health technologies). We have discussed the examples of the display assembly and lithium rechargeable battery in this regard. Therefore, the argument for protecting the manufacturing of complex technical systems is just not economic but a question of defense.

But if the government does decide to develop a national policy of emphasizing manufacturing, it has to create a public-private partnership and for a variety of reasons. Most of the basic and much of the applied research are conducted in the public sector, such as the national laboratories or the universities, whereas most of the commercial applications are made in the private sector, whether the large industrial laboratories or the small high-tech firms. Thus, the public and the private sector must cooperate to overcome the valley of death where it exists and to accelerate the rate of product development. A rapid rate of development of radical product and process innovations would require a particularly tight connection in the idea innovation network, as I have already argued.

Another reason for the need for cooperation is the growing complexity of products and the technical services attached to them. Elaborate supply chains for each of the technological sophisticated components have attached to them separate idea innovation networks. So, this complex technical system is difficult to coordinate by market principles, and thus requires a joint public and private sector coordination.

But finally, the major reason for cooperation is the difficulty of achieving large volumes of production in manufacturing with relatively sophisticated machines that compensate for the lower wages elsewhere, as I observed in the introduction to this chapter.[11] Public-private cooperation in each of the three potential strategies would greatly increase the possibility that large volumes with low prices could be achieved. The first strategy that I suggested was a radical increase in flexible manufacturing. What is needed is reconfigurable and agile manufacturing systems that respond to ever shorter innovation cycles and rapidly adapt to customer demands for new products and new product features. The whole objective of Chapter 1 was to make clear exactly that these new product features should include greater customizations, higher technological sophistication, and better design. Also, in the same chapter, we discussed the many means of quality, which is essential for success in the marketplace. Radical reductions in the use of energy and scarce materials would also be a major plus. But these kinds of manufacturing systems can only be created via cooperation between the public and the private sectors.

The second strategy is to focus on generic platform technologies. For example, a new logic switch in nanoelectronics would include the following:[12]

- Concepts of new circuit design technologies based on the properties of individual molecules
- Generic fabrication methods for radically new classes of materials with unique electronic properties
- Generic methods for inducing novel compounds to self-assemble into the precise structures needed for new electronic devices and architectures
- Generic methods for interconnecting new devices into circuits

In turn, development for this example in nanoelectronics also supposes the creation of new measurement capabilities at the molecule level, including their shapes, dimensions, and electrical characteristics; techniques for manipulating individual electrons; and of course, the accumulation of more data on the physical behavior and reliability of new nanoelectronic materials.[13]

Finally, the third strategy is to attempt to create the third industrial divide along of the lines of Modumetal reported in Chapter 1. It is precisely in the pursuit of this third strategy where the greatest cooperation is necessary because just as with our example in nanoelectronics, one can perceive that the development of the third divide requires the cooperation of a number of disciplines, including physics, materials, engineering, and information technology, to achieve this objective. Obviously, it would be best if both the private and the public sectors pursued all three strategies because their relevance and possibility of achievement differs by sector.

Despite these many arguments for cooperation between the public and the private sectors, for many this idea at the team level seems like a radical step away from free market principles. In the free market system, the usual assumption is that these two sectors should be at "arm's length." Yet, if we are to compete against other countries where there is a considerable amount of cooperation and especially when the pace of technological change becomes a decisive factor, we are placing ourselves at a considerable disadvantage.

There are really three fundamental arguments against this concern: one is historic; another is the present existence of this arrangement; and as I have already noted at various points, the third is the nature of the national competition that presently faces the United States. The historic argument is that free market principles worked reasonably well in the early stages of industrialization when the closed model of industrial research laboratories at GE, AT&T, GM, DuPont, P&G, and RCA had plenty of time to slowly develop their radical innovations, and most critically the United States was the only country that was creating them, with perhaps the notable exception of Germany. None of these conditions prevail now. Fewer than half of the twenty largest firms in terms of their expenditures on RDT are now located in the United States, whereas previously most were. Meanwhile, the speed of radical innovation development has accelerated, and most critically, the globalization of RDT expenditures and the national industrial policies of many countries have made the issue of the competition more global, as we saw in Chapter 1.

A corollary argument is one that we have already visited, namely, the opposition to any idea of picking winners and losers. As I have already suggested in Chapter 1, there are some risks in this, but attempting to facilitate entire sectors with various policies reduces the risk somewhat, as does knowing some of the directions of evolutionary change. The various suggestions about generic platform technologies and the creation of the third industrial divide exactly fit this pattern of helping many sectors and not just a few firms.

In the next section, I document the many ways in which public-private cooperation is occurring and, more particularly, how much in the past it has been a vital part of the building of America. Perhaps the most telling example, because it is not defense-related, is how much our agricultural extension system created the most bountiful food production system in the world. As indicated in the next section, it was built on other than free market principles, including close cooperation between the public and private sectors and, most importantly, network coordination. With this observation, let us now examine the various examples of public and private cooperation.

In summary, a major obstacle to overcome is the belief in free market principles as the best way of generating radically new products and processes. One way of overcoming this obstacle is to recognize in which ways public-private cooperation of various kinds is being built, our next topic. As I review the evidence, it will become apparent that the United States has moved a good distance toward cooperation, leaving two remaining questions to think about: whether it is the correct approach and whether much more could be done.

RETHINKING HOW THE PUBLIC AND
THE PRIVATE SECTORS CAN COOPERATE

In rethinking the current extent of cooperation, it is useful to examine in what forms it exists and whether it can be improved. As I have already noted, the Department of Defense (DOD) and, more recently, the Defense Advanced Research Projects Agency (DARPA) have been very successful in sustaining the innovative edge of this country in those sectors deemed vital to national defense. Indeed, without the DOD having supported the development of computers and the Internet, the state of the American economy would have been much worse much earlier.

There are two kinds of support that the federal government outside of the DOD gives to industry:

- Funding basic and applied research in the national research laboratories and universities, which in turn provide outreach of various kinds
- Funding directly various firms that have innovative proposals

After discussing these two ways of stimulating radical innovations that can be commercialized, I want to focus on those sectors of the economy that do not receive much support and where many opportunities exist for restoring the innovative edge if the right kind of cooperation is created.

At the beginning of Chapter 4, several studies criticized the national laboratories for having a stovepipe mentality and not commercializing their discoveries fast enough to be useful. This assessment must be considered against the backdrop of the impact of the national research laboratories on radical product and process innovation in this country. At various points, the one hundred annual awards made by *R&D Magazine* for commercialized innovations have been cited. Accepting the limitations of this measure, the number of awards made either along or in collaboration with another organization, including supported spinoffs, averaged about 13 percent in the decade of the 1970s and rose to over 50 percent from 1991 to 2006.[14] In other words, for the sectors outside of computers and pharmaceuticals, the national laboratories would appear to be having a measurable impact on the launching of commercial products. It should be remembered that in this same period the large industrial laboratories have largely disappeared, as was reported in the Introduction. Of course, the largest limitation is the fact that 100 are selected each year, implying a steady state in the rate of product innovation, which is anything but true. The awards from 2002 to 2006 could be much less radical than those of the 1970s. And conversely, one might also argue that it is harder to win in the most recent period because of the large number of innovations.

Some of the national research laboratories have been encouraged more than others to develop good working relationships with businesses. As an example, the National Renewable Energy Laboratory (NREL) has had a long-term relationship with a small tech company, Abound Solar.[15] The invention process started in 1993 when the three founders were faculty members at Colorado State University and began working with NREL scientists. In 2007, the invention process had moved far enough along that it was placed in the Photovoltaic Technology Incubator Program. The goal of this program is to accelerate the rate of movement from prototype to full-scale manufacturing. The company has developed a process to deposit semiconductor layers very rapidly—in a matter of minutes. The specific manufacturing research problem was how to take a 4-foot by 2-foot piece of glass and turn it into a solar panel in two hours at a cost of less than $1 per watt. By 2009, the manufacturing problems had been solved and Abound Solar had moved from 33 employees to 230 and received substantial venture capital funding. Admittedly, this is a success story and does not represent an assessment of all that NREL has accomplished or failed to accomplish.

Some of the recent efforts at facilitating cooperation between the public and the private sectors at Oak Ridge National Laboratory can be cited.[16] A BioEnergy Science Center was designed to emulate a start-up and develop cellulosic ethanol, tapping the scientific expertise to be found in the national laboratories, academia, and private firms. Another effort of this national lab is the building of an office park on its campus to attract small high-tech companies that would facilitate more exchanges with the scientists at Oak Ridge.

One of the untold and underappreciated aspects of the national laboratories is their development of technological instruments that can be used by private business. Generally, about three-fourths of the users are academics, and there is no assessment of how much impact the remaining fourth has had on innovation in this country. Besides the BioEnergy example just given, are the four NanoCenters that the Department of Energy (DOE), each at a different national laboratory, has created recently as research facilities. Some of these new scientific technologies for research are impressive. The National Synchrotron Light Source at Brookhaven is one of the most widely used research facilities in the world because it is a giant microscope for measuring electronic and atomic structures of matter and for studying the surfaces at the atomic level. A new facility to upgrade this technology is presently being built and will become available in 2015; it will have ten thousand times the amount of light of the previous one and will allow the probing of nanomaterials with considerable precision.[17]

How does one reconcile these various signs of success with the criticism about the slowness of development that has been reported? The answer may be that both are correct. Despite the many outreach programs, including incubators, centers, technological instruments, and even joint research projects such as the one at NREL, it is entirely possible that more scientific discoveries could be converted into commercial products—the valley of death does exist—and the speed of development could be considerably increased if the cooperation consisted of joint research teams as suggested in Chapter 2, where both the public and the private sectors had researchers experienced in manufacturing and quality research. So far the logic of the national laboratories has focused on technology transfer rather than solving the many problems in *manufacturing and quality research* that have to be resolved before a product can be successfully launched. Thus, in rethinking the extent of cooperation between the public and the private sectors at the level of the national research laboratories, there is a need for research on alternative methods for accelerating the rate of product and process innovation. The fact that some of the national laboratories are exploring better arrangements provides a wonderful opportunity for policy research to assess which methods can most effectively accelerate the range of radical innovations being commercialized.

Quite different from the national research laboratories are the programs in which the U.S. government invests directly, but not always, in businesses and usually, but not always, small companies. One example is the Small Business Innovation Research (SBIR) Program. Federal agencies with large research budgets are required to devote 2.5 percent of their RDT budget to supporting firms with five hundred employees or fewer.[18] For phase I projects, the grant is $100,000; for phase II projects, it is $750,000. In 2004, the SBIR Program resulted in a disbursement of $2 billion to some 6,300 firms. Starting in 1982, one of the R&D 100 Awards went to a firm supported in this program; the number jumped to eleven in 1991, and since 2002 between 20 and 25 percent of the awards have gone to firms supported by the program.[19]

One limitation of the SBIR Program is that it provides support for only three years. Thus, the question arises of how to continue funding for those projects that have reached some level of promise but are not ready to be commercialized. Again, as usual, DOD and the intelligence agencies have led the way. The Central Intelligence Agency has now established a venture capital fund to help move the most promising discoveries of the projects it has invested in into commercialization, In-Q-Tel.[20] One of the ninety projects in its portfolio is the Modumetal project, described in Chapter 1. The U.S. Army is also developing its own venture capital fund, as has DOE in combination with Battelle. In each instance, the venture capital funds are nonprofit and are paid back when the firms that they invested in become successful so that the fund can continue.

Another program, which was discontinued in 2006 because some felt it was against the principle that the private sector should pay for its own research but which has recently been restarted, is the Advanced Technology Program (ATP) of the National Institute for Standards and Technology (NIST). In contrast to many other programs, a considerable amount of evaluation of this program has occurred, precisely because it has been controversial. The objective of ATP was to accelerate the development of innovative technologies that were considered to be of high risk and unlikely to receive funding from other sources. In fifteen years of operation, the program disbursed about $2.3 billion to those projects that met these criteria and presented both a good research plan and a business plan on how to commercialize the product.[21] Each project had to contribute part of the funding, with the amounts varying somewhat but usually being close to a fifty-fifty match. The range of projects was broad and included advanced materials and chemistry, discrete manufacturing studies, biotech, electronics and photonics, and information technology.

A total of 768 projects had been funded by 2004, when the program was in the process of being phased out. Of these, 630 went to firms with fewer than five hundred employees, and one-half of these went to firms of fewer than twenty employees. In other words, the small high-tech company was the main beneficiary. Although complete data were not collected—the first three years is missing—the 768 ATP projects produced 1,700 publications, indicating the contribution to science; 1,418 patents were accepted and another 1,500 were pending, indicating the contribution to technology; and about one-half of them launched a new product, indicating the contribution to the economy. Perhaps the most vital indicator is that 62 percent of the projects attracted additional funding so that the ATP investment had a multiplier effect on encouraging investments in high-risk innovations. Yet, another indicator of a multiplier effect is that patents that had been filed had been cited in another 12,000 patents, showing that the technology developed has been spreading.

In evaluating the program, case studies of the first 150 projects completed were made to assess the extent of their success or failure in more detail. About 31 percent were deemed to have limited success, which corresponds fairly closely to the overall patterns just reported about papers, patents, and commercialization. In contrast, relative to how innovations create employment, about one-third of these projects doubled their employment and another 14 percent grew by 1,000 percent.

What are some examples of the kind of projects that were supported and were highly successful?[22] X-Ray Optical Systems of Albany, New York, developed a high-energy imaging technology that reflects X-rays and neutrons through thousands of tiny glass tubs. Seven patents have been received and *R&D Magazine* recognized the accomplishment. The number of employees grew from one to twenty.

The importance of developing new manufacturing processes is demonstrated in the process developed by Displaytech of Longmont, Colorado. It was able to mass-produce ferroelectric liquid crystals (FLCs) that increased the quality of display image by 600 percent, increased the lifetime of the product by 100 percent, and reduced unit costs of manufacturing from $6,000 to $160. The number of employees has grown from 20 to 150, three patents have been awarded, and it has been widely introduced into high-definition televisions (HDTVs) by various companies.

Another example of how processing time was radically reduced is the development of DNA diagnostics on a chip by Orchid Bioscience. Previously, DNA analysis required two technicians with several machines and cost about $100,000. Once the chip was developed, processing time was reduced from four weeks to one week and costs by 70 percent. Five patents were awarded, and growth in sales went from $1 million a year to $62 million in 2004.

Another project given to Benchmarking Partners in Cambridge, Massachusetts, developed software for coordination of the supply chain that had a considerable impact on increasing sales and reducing the costs of inventory in the supply chains of both a food store and an apparel company.

Unlike the SBIR Program, ATP grants were given to large companies as well, perhaps the most controversial aspect of this program. GE Global Research of Schenectady, New York, developed a radical advance using amorphous silicon panels to detect heart disease and breast cancer. Genex Technologies of Kensington, Maryland, is developing a revolutionary facial recognition technology.

Both of these programs would appear to be largely successful. The question still remains whether both could be made more effective if two stages were recognized: first, the development of the advanced technology or prototype, and second, the development of the manufacturing and quality control processes. In the first stage, perhaps matching between the national laboratories and the small firms would facilitate the rate and speed of radical product and process innovations. In the second stage, perhaps matching between the small firms and manufacturers that have more technical expertise would reduce the costs and improve the quality. In other words, both programs perceive the issue of providing initial funding and expecting venture capital to provide for further development, and not so much as how to coordinate the relevant expertise and specific competencies in manufacturing and quality control. Clearly, in some of the examples manufacturing problems were solved so the issue of expertise is not always relevant. Beyond this, a number of projects in the ATP involved partners of various kinds but obviously ones that the applicants found. The question remains whether a better matching process could be developed.

Could the U.S. government develop a registry of competencies along the lines that P&G has done for its specific businesses, but do this for each of the six arenas of research in each of at least the high-tech sectors of the economy as well as important noneconomic sectors? This registry, which would have to be updated continuously because of the evolution in science and technology, might help overcome the market failures in the matching process. Again, the same question remains whether more attention to the development of partnerships between public sector and private sector researchers might accelerate both the rate and speed of radical product and process innovations development.

Another issue in rethinking the extent of public-private cooperation is to set a five-year limit on the licensing of any product, process, or patent to any firms except those that guarantee that the knowledge will be used in the United States and especially insofar as it impacts on employment. This restriction is necessary for several reasons. First, as suggested, various components are parts of complex technical systems that can overlap various sectors. If we want to gain the rewards of our radical innovations for our society and especially for increasing employment from investments by the federal government, we need to ensure that the knowledge is used for manufacturing within this country. Second, as has been argued in various places, the United States is competing against countries that do think about their competitiveness in collective terms, and not always because of government coordination. In particular countries, the reasons for retaining the technology vary: In some, it is the influence of the family;[23] in others, it is the influence of associations of various kinds.[24] But regardless of the origin, the fact remains that we need more of our innovators to think about the problem of American competitiveness.

Still another issue in rethinking the cooperation between the public and the private sectors in programs such as SBIR and ATP is the question: What is next? What is the second act? As was noted in the famous research on hard disk drives, each company maintained a static strategy and was replaced by the next company that had developed a smaller model with a larger memory, a technological development that should have been obvious.[25] The reality is that many companies rest on their laurels or cash in quickly on their patents rather than realize that one must continue to run in order to stay in the same place. One advantage of having a five-year limit on cashing in quickly on various technical advances is that it may encourage small firms to continue to advance their technology and think about the competitiveness of the United States as a collective goal.

Although the primary focus in this book is on the high-tech sectors that are presently declining, I do not want to ignore traditional industries that have been more strongly impacted by imports. Some of the current scientific objectives such as nanotechnologies and materials science can impact on so many different products; they afford opportunities to produce novel new products in many low- and medium-tech sectors. For example, one area is the development of synthetic materials to replace wood, but with many of the properties of wood, including color and texture.

When we shift to the small- and mid-size manufacturing firms, some 282,000 of them, many of which do not have their own R&D departments and therefore little capacity for scanning the scientific and technological advances that are occurring, we need a different strategy than one that is being used for small high-tech companies. A recent study of these firms found that fully a third of them, or about 90,000, were defective in corporate strategy of all kinds.[26] Yet, these same manufacturing firms feed into many of the components involved in various complex products; they are part or could be part of these technical systems that we want to protect. As we have seen in Chapter 5, Apple assembled many of its components from Asia. What is needed is to provide Apple with good alternatives in this country if we are to save our workforce and standard of living.

How to do this, especially given that many of these 282,000 firms are not necessarily located in a single region? We need a national system but one that is geared to the quite dramatic differences that exist between sectors, both economic and noneconomic. There is a model, one developed within the United States a century ago, that can provide the answer to this question.

This model can be found in agriculture, perhaps one of the most extreme examples of multiple regional systems built within a national system. The different regional systems are based on distinctive crops, livestock, and types of fishing. The nature of the food product may be more concentrated in certain states than others, but the key point is that usually a single crop such as corn, or wheat, or livestock such as cattle or pigs covers a large area (e.g., several states).

Even though agricultural production now presents a relatively small proportion of the nation's GDP, again important historical lessons can be gleaned from examining what was done.[27] The agricultural extension services of the U.S. Department of Agriculture (USDA) created long-term and consistent economic growth rates in agriculture. The USDA and in particular its extension services recognized systemic problems in the production of food and then attempted to solve them. What is perhaps most critical is that they in effect viewed the food production system in light of the idea innovation network, focusing on basic research, applied research, product development, manufacturing research, quality research, and commercialization research. In addition, they viewed the system in light of both economic and social problems that had to be solved.

Over the last century, American agriculture has been a continued success story, so much so that until recently the industry has been dominated by the problem of overproduction due to the steady stream of innovations in plants, animals, machines, and farm management. Some of the major innovations are soil conservation techniques, hybrid corn, specialized machines for planting corn, integrated pest management, and more recently, the adoption of the computer for farm management. The USDA did the basic research supplemented by the extramural program of research grants to the land-grant colleges, which focused on applied research and product development. The key component, however, in the steady growth of agricultural production was the disseminating role of the extension agent who learned new techniques developed elsewhere and passed them on to the farmers in his area. For example, in Iowa during the 1930s, the agents quickly disseminated better varieties of oats and alfalfa when they became available. They also conducted research on the quality of the soil to measure the amount of lime. Another thrust of their work was to improve the quality of egg and of lamb production, establishing grading systems that protected the consumer. Thus, agricultural innovations diffused relatively quickly throughout the various farming states. The combining of these innovations (e.g., nitrogen fertilizer, hybrid seed, conservation practices, and the effective control of pests and disease), each of which had only about a 10 percent impact, together raised the yield of corn from 15 to 20 bushels per acre to 80 bushels in North Carolina.[28] In these examples, we also see the attention given to the different arenas in the idea innovation network. These improvements were replicated in every major crop and livestock category in the United States, resulting in the tremendous surpluses that created the depression in the industry during the 1980s and 1990s. In addition, more recently, the focus has shifted to new kinds of products such as kiwis and ostrich production (low in cholesterol) and to new kinds of farming (less use of water and of energy). Given the limit in the amount of growing area, perhaps the biggest tension presently is between the need for grasses for ethanol fuel and the need to maintain food production, unless new kinds of grasses can be grown in the desert as already suggested.

What has not been well appreciated in the economic history of the United States is that the USDA continued to change its context by tackling new problems and went considerably beyond the simple mandate of increasing productivity on the farm, ranch, or orchard. The full range of problems associated with life on the farm included confronting family issues such as the role of women, leadership training for youth in the 4H clubs, the economics of farming, provision of credit, and community development. All of these efforts led to the creation of a considerable amount of cooperative behavior in these states, with rich rewards for the United States in other ways.[29]

Equally unappreciated in the economic history of the United States is that this system is the exact opposite of market competition. The objectives of the USDA, the land-grant colleges, extension agents, and specialists—which constitute a very elaborate network of organizations, considering the farm as an organization as well—were, in fact, to buffer the farmer from the vagaries of market competition. This is not to say that each farm with a particular food product did not have a considerable amount of competition but only to say that they were encased within a network that was trying to solve their problems. Paying farms not to grow crops, storing surpluses, and maintaining floors on prices are all the more extreme examples of protection against the market that has been provided by the federal government at various times.

As each decade produced a new set of problems, the extension services quickly recognized them and grappled with them, from the problems of shortages of labor during the Second World War to the problems of overproduction in the 1980s and 1990s. The USDA also made a concerted attack against poverty during the Great Depression and especially tried to change the tenant-farmer situation of poor blacks, but was prevented from doing so by southern politicians that controlled the main levers of power in Congress.

What are some potential historical lessons for the United States and the current discussions of American competitiveness? First, the Department of Commerce should be renamed the Department of *Industry* and Commerce. This would call attention to the need for an industrial policy beyond the simple notion of free trade (or its opposite, tariffs or certain kinds of trade restrictions) and, more specifically, the importance of innovation for positive trade balances. As with the model of the agricultural service, the idea is not to pick "winners and losers"—the usual complaints made against industrial policy—but to create the right conditions so that there are multiple winners and losers just as there were with the consequences of farm policy. Not all farms succeeded, and one could argue that many of policies gave preference to the corporate farms because the implication of many of the practices was to routinize and industrialize the production of food.

Second, the focus of the effort would be to construct a system of research in each of the six arenas emphasized in the idea innovation network—basic research, applied research, product development, manufacturing research, quality research, and commercialization research. As indicated in Chapter 1, the strategy would be to get ahead of the innovation curve in each of these low- and medium-tech sectors by emphasizing customization and reducing environmental impacts, that is, using less energy consumption or scarce natural resources and lowering health risks.

Third, as part of this research system, different organizations would be created or involved in order to specialize in the specific research arenas. The Department of Industry and Commerce would build national centers of technology for each industrial sector and locate them in different parts of the country. A good model is the National Forest Products Laboratory located in Madison, Wisconsin. But there are many models for these dedicated technology centers to be found in a number of different European and Asian countries, as I have alluded to in some of the examples of creating Silicon Valleys. These centers would concentrate on basic and applied research, but in addition would conduct an extramural program of research to be established in the different state universities and even community colleges if they offered a strong technical program that was relevant, as many do. Manufacturing and quality research might be conducted in special experimental programs with firms. Indeed, a partnership between the public and private sectors is essential for this initiative to succeed.

Fourth, the key component remains the extension agent. Individuals with experience in the industrial sector should be hired and trained. They become the vital link in the coordination process of determining what problems require solutions and then disseminating the solutions to these problems to all firms in the same sector. Again, the idea is not to pick winners and losers but to allow firms to profit from the improved product capabilities, manufacturing techniques, and quality procedures that would provide a marketing edge.

Fifth, the range of problems should not be simply scientific and technological but social as well. In particular, the upgrading of education of both workers and managers should be a high priority. In particular, this might have some positive repercussions on boys staying in high school longer if good technical programs were offered that led to advanced technical training in community colleges similar to the educational systems of Germany and Italy. Creating applied research engineers and scientists interested in the problems of a specific industrial sector should be another. And finally, some attention should be given to how to build strong manager-worker cooperation and stronger communities in which these businesses are located.

Sixth, special attention should be paid to the potential of restarting our machine industry by building the tools and techniques that reduce energy consumption, provide much higher quality in the different senses in which I have defined it, and reduce health risks in the low- and medium-tech industrial sectors that can be sold overseas. Both Italy and Germany have captured considerable exports of machines and in the different sectors of the economy. The key point is that one needs machine tools for each of the different sectors that are customized to fit the technological and cultural patterns of each country.

The immediate issue is how politically viable this is. The success of the USDA was built on the coalition of a large number of states interested in receiving the benefits of this program. The same broad political alliance is possible in this context because it would be to restore American competitiveness and protect America's standard of living. It would not be picking winners but instead be designed to stimulate radical product and process innovations in many sectors that are presently low- or medium-tech.

In summary, a number of ideas have been suggested about how to rethink cooperation between the public and private sectors. It has to move beyond simply the idea of funding promising ideas to cooperating in the development of competencies and especially in the arenas of manufacturing and quality research. In addition, American CEOs of large or small firms have to—as Immelt has said—think about improving the competitiveness of this country.[30] In these general guidelines are ample opportunities to create different arrangements with each sector.

THE NEED FOR MORE RESEARCH ON THE EVOLUTION AND ITS FAILURE

At the conclusion of the Introduction, I suggested that there is a need for a new policy model based on the evolution of science and technology. The first implication of this idea is the rethinking of the science and engineering indicators that are collected by the NSF and reported every two years online.[31] The current set of indicators emphasizes the investments in basic research, applied research, and product development. But if one accepts the model of evolution in which there are six arenas, then one needs to measure the amount of investment in manufacturing research, quality research, and commercialization research as well. Collecting data on these investments in what I called BAPMQC (basic research, applied research product development, manufacturing research, quality research, and commercialization research) in Chapter 1 can make more apparent particular arenas that might have been neglected. For example, I have suggested that manufacturing and quality research have not received the kind of emphasis that they deserve. But to be sure, one needs evidence about this. In addition, the collection of data should recognize some of the new implied sectors, particularly noneconomic ones such as homeland security and reduction in global warming. Each time the government establishes a goal, especially if it is likely to be long lasting, then it should become designated as another sector in which scientific and technological data is collected.

Parallel to the broadening of measures of investments, measuring the number of scientists, engineers, and technologists who are concentrating on research in one or more of these six arenas should be done. Earlier in this chapter, I suggested a further refinement, namely, the measurement of competencies. This builds on an old idea of the Dictionary of Occupational Skills, which was designed to create a common definition for occupations across sectors. This proposal moves to a finer grade to identify specific competencies within particular arenas of a sector. Perhaps the construction of such a system is too expensive for the Department of Labor or the NSF to undertake, but if GE could identify competencies for its research project on the gasification of coal and if P&G has been able to do the same across its various products, then perhaps it is not beyond the realm of possibility if done cooperatively with a random sample of firms. A bigger limitation is how quickly it can become obsolete since the pace of research findings and, by implication, the irrelevance of various competencies are accelerating.

Presently, the NSF measures a number of performances and at the sector level. For example, in the Introduction, I reported the trade balances in the high-tech sectors, one clear economic measure. In addition, I would add the following measures to assess how much radical innovation is occurring at this level:

- The degree of radicalness on the various dimensions of the product mix
- The average speed of product development or time to market
- The competitive position of the product mix

It goes without saying that the level of analysis in this context is not the single research organization but all of those organizations producing products or providing services in a particular technological sector, including the noneconomic ones such as defense, education, and global warming.

Consistent with the repeated emphasis on research on qualities and research on the customization of manufacturing, product mixes at the sector level should be evaluated along a number of different attributes or dimensions to understand correctly the competitive position of the sector vis-à-vis other nations. Such data would provide policy makers as well as the leaders of various research organizations with a sense of what opportunities are available, especially if they shift their portfolio to emphasize attributes or dimensions that have been ignored.

How easy is it to determine the competitive position of the product *mix* for policy makers? Surprisingly, this is the easiest part of the data collection. There are trade journals that provide long lists of various dimensions on which the product mix can be assessed. It is the same as benchmarking that we discussed in Chapter 5. Marketing executives are acutely aware of the relative strengths and weaknesses of their products in a global context. So this information is easy to collect for products, if not for services. What is more difficult is measuring the extent of improvement on various hidden costs of the product mix, that is, how "green" are both the product once in use and its manufacturing, a dimension that has increasingly been recognized as important as the characteristics of the product or service itself. But as these qualities of product become more and more critical for national goals, this kind of information will also be easier to assemble.

The average speed of development has become a critical factor in maintaining a country's competitive position in the high-tech sectors. We have already discussed this in the context of the automobile industry in the Introduction and especially in the management of the supply chain in Chapter 5. Nor are automobiles the only technological sector where the speed of product development is as important as the radicalness of the product. Semiconductor chips and drugs are two other examples, even if the former seems to be a race of rabbits and the latter a race of turtles. What is interesting to observe is that it is precisely in those technological sectors where speed is most critical that most of the RDT is spent and where both the radicalness of the product and speed are critical factors in successful commercialization of the product. Therefore, NSF should examine the time to market as part of the feedback on a specific technological sector.

The third standard, namely, the financial success of the product in trade balances, is easily measured, as we have seen in the Introduction, but not that of the *mix, that is, what the trade balances of specific kinds of products reflecting different product mixes are.* The issue is, why does a particular sector do well? Is it because of its quality or because of a specific attribute such as ease of use? Also given the spread of foreign ownership, some of the positive trade balance may be a consequence of foreign-owned firms. As the evolutionary process toward more and more internationalization of each high-tech idea innovation network occurs—as has happened in pharmaceuticals, telecommunications, and semiconductors—the more difficult it is to disentangle the policy effects on that specific sector.

A major implication of these performance measures is to shift away from productivity measures to ones that attempt to establish how much radical product and process innovation is occurring in each sector. The objective of the performance measures is to determine whether the United States is falling behind in a sector, and if so, whether the reason is in part because of the lack of investment or technical expertise in one or more arenas. At various points, the argument has been made that if evolution is proceeding correctly, then the United States would continue to have strong innovative performances in each of its high-tech sectors. This supposition provides a basic framework for beginning to identify potential obstacles at the meso level, the next major implication of the policy evolutionary model.

The second major set of data collection implied by the evolutionary policy model presented in Figure 1.3 is collecting data on the evolutionary patterns predicted by the model:

- Differentiation of research arenas with separate organizations
- Creation of new research organizations, frequently small ones, within some of them
- Formation of networks within and between the differentiated research arenas
- Strengthening of the connections in these networks
- Extent of radical innovation in each of the six arenas

The first data question in the framework is the following: Are there separate arenas as reflected in research organizations that have been differentiated? How much each of the six arenas is differentiated indicates how much focus there is likely to be on a specific arena such as manufacturing research or product development. This measures the extent of the differentiation, which is predicted from the growth in RDT investments. Admittedly, knowing how much differentiation should have occurred is more difficult to assess. Two ways of benchmarking the extent of differentiation are possible. One is an international comparison. For example, German policy makers observed the development of biotech firms in the United States and the emergence of networks between them and the pharmaceutical companies, and noticed their absence in Germany. In Chapter 5, in the discussion of a fourfold typology describing the differences between sectors, I observed that certain sectors in other countries have accumulated more research and practical experience that may well be expressed in greater differentiation of the arenas. These could be used as benchmarks for the relevant sectors in this society. Another is the comparison with a similar sector (in terms of both industrial structure and RDT expenditures; see Chapter 5). Given similar expenditures and differential development, the slower of the two is probably demonstrating signs of evolutionary failure.

The second data question in this framework is the following: How many new research organizations, especially small ones, are there? The answer is perhaps a telling sign of whether we are creating new competencies within a specific arena and sector. In particular, federal programs such as SBIR and ATP should try to encourage the development of small high-tech companies in certain sectors if deemed important for the growth of the society.

The third data question in the framework is the following: Are there networks connecting teams within arenas and between them in the idea innovation network? Whether these research teams are connected impacts on the trade balances. If two of the arenas are not connected, then a valley of death has emerged. In the past, when all arenas were within the same organization, such as P&G, DuPont, and GE (i.e., the closed innovation model), the issue of connectedness did not seem to be a problem. Yet, gradually over time, even— within the same organization—connectedness between research teams became problematic, as we have seen in the case studies of IBM and the Pasteur Institute, as well as the general discussion of stovepipes. The disconnectedness that occurred between Bell Laboratories and AT&T is well known, as it is in Xerox.

The fourth data question is the following: How strong is the connection? The argument advanced in Chapter 1 is that as the radicalness of the technical achievement in a specific arena increases, frequent and intense interaction in the network within arenas and between them has to increase to understand the hidden assumptions and intuitive reasons behind the findings. The usual measures of links between arenas, papers, and patents have several problems with them. Generally, they occur too late for any effective policy intervention, nor does it mean that there has been effective communication. Instead, the following are useful measures of the strength of the connection:

- Transfer of people from one research group to another, both within and among organizations
- Joint research projects involving face-to-face collaboration among researchers, as distinct from long-distance collaboration
- Joint paper-writing
- Strength of managerial, financial, and research ties among organizations in joint ventures
- Strength of ties among actors in research consortia[32]

In Chapter 5, the case of SEMATECH illustrates how to measure the strength of the ties by the very elaborate decision-making structure that was created.

Table 6.1. Metrics of radicalness of output for each research arena in the idea innovation network

Functional Arena	Measures of Radical Advances in Output
Basic research	Percent increase in the modeling of some scientific behavior. Solution to a central problem. Identification of new concepts or processes.
Applied research	Percent increase in control over some desired attribute.
Product development or product innovation	Percent increase in different performance characteristics weighted by their importance. Addition of new properties to the functionality of the product.
Production research or process innovation	Percent increase in productivity. Percent increase in ability to customize production. Percent decrease in energy consumption.
Quality control research and research on qualities	Percent decrease in defects. Percent decrease in operating costs. Percent decrease in various externalities weighted by their importance.
Commercialization research	Percent increase in customer satisfaction. Percent decrease in delivery time.

At various times, I suggested that the meso level offers a particular vantage point that helps in providing assessment on the competitiveness of the United States. A good example of this is that each of the six arenas represents a different kind of research that can be used to assess the radical innovation in that sector. Table 6.1 lists some suggested measures for each of these six areas.

One advantage of focusing on radical innovation in each arena is to provide quicker assessments of where gaps might be developing.[33] One major advantage of measuring the technical outputs of all six arenas is that it allows the evaluator to provide feedback on how the entire network is functioning, identify problems at the sector level, and explain why the United States may not have strong trade balances in certain economic sectors or is failing to achieve certain noneconomic policy objectives such as those in global warming, national security, or health.[34] The data that are collected suggest a series of obstacles that could explain the lack of innovation in a particular arena. But there can be other reasons at the organizational level, which would then call for an in-depth study of one or more research organizations in the arena, a topic for Chapter 7.

Although Table 6.1 provides suggestions for each of the six arenas, it may not be necessary to measure all six. This is a question of how much functional differentiation has occurred in the idea innovation network, that is, how far the evolution of knowledge growth has progressed. For example, if basic, applied, and product development research are combined in a biotech company, and the manufacturing, quality, and marketing research are combined in a pharmaceutical, one can concentrate on the technical outputs of product development of the biotech companies and the product outputs of the pharmaceutical firms.

This assessment is only a first approximation and, again, it may be necessary to examine the outputs of the other arenas within the biotech companies and the pharmaceutical firms because the blockages can be organizational, with bottlenecks between basic and applied research in the biotech companies or between manufacturing research and quality research in the pharmaceutical companies. As more and more of the arenas become functionally differentiated, one is forced to measure the technical outputs of each arena. This may seem like an enormous amount of effort, but it speaks to the issue of understanding the innovation processes at the level of the idea innovation network and provides a number of advantages for restoring the innovative edge.

The radicalness of the advance in the technical output is a question of the context and how a radical advance is defined. As was indicated, the indicators have to be specific to the particular discipline or product. Generally, increases of 1 or 2 percent in some output would be incremental. Radical advances certainly do occur and reflect a larger percentage increase in the specific percentage defining the achievement in an output. Automation that improved the throughput of dishwashing machines by 300 percent clearly represented a radical advance as a consequence of manufacturing research. The data collected in the foregoing helps us to identify the presence of failed evolution.

The third major implication of the new policy model is to determine why evolution has failed. This is a much more difficult task. The list of obstacles can be a useful starting point, but probably the NSF could collect data on only a few of these obstacles and those suggested by the measures used earlier in this section: not enough funding of manufacturing research, weak connections between one or more arenas in a sector, and the lack of technical competencies in a specific domain. But some of the obstacles and, more importantly, the blockages behind this are beyond the capabilities of the NSF to collect data on a regular basis; they required detailed policy studies. Therefore, the next chapter shifts to this important topic of feedback on blockages and especially at the organizational and team levels.

In summary, without good data collection on relevant indicators, we remain ignorant of fundamental policy problems. Knowledge gained from data collection is the advantage of this new policy model based on evolution and failed evolution. It opens our eyes to new policy issues that need to be discussed and debated and eventually resolved.

EVOLUTION AND THE NEED FOR MORE PUBLIC AND PRIVATE COOPERATION

The central argument in this chapter is that we need to rethink our industrial policy in light of the growing complexity of the technical systems associated with products and services and especially in the high-tech sectors. The focus of this new policy should be on manufacturing because not enough high-paying service jobs are being created to replace the manufacturing employment that we are losing. In particular, the exporting of R&D research overseas means the loss of the most critical component in the idea innovation network and, one might add, in the supply chains as well.

The evolutionary processes, spurred in part by concerns of the DOD about protecting national security, toward greater cooperation is more advanced than most are aware of. However, two ingredients in this cooperation appear to be missing. The first is an adequate coordination of the idea innovation network within each sector. The second is equal participation in the research teams, especially involving competencies of manufacturing and quality research. In particular, the small- and medium-sized low-tech firms have largely been ignored in the various efforts of the U.S. government, and yet they hold many opportunities for getting ahead of the innovation curve.

But while detailed information about coordination of these networks and whether or not failed evolution is occurring is lacking, the NSF has concentrated its emphasis on the investments and capabilities primarily in the arenas of basic research, applied research, and product development rather than in the other three arenas of the idea innovation network. Thus part of the cooperative efforts between the public and the private sectors should include the collection of more data so that informed decisions can be made and failed evolution or the valley of death can be detected more quickly.

Throughout this chapter are four consistent policy recommendations for the federal government:

1. Create public and private sector coordination of complex technical systems, especially in the high-tech sectors.
2. Emphasize more manufacturing and quality research in the cooperation between the public and private sectors.
3. Create extension services and Silicon Valleys, particularly for the sectors with many small- and medium-sized firms.
4. Expand the data collection of the federal government to include much more data on the sector level, such as technical progress in the arenas and even a registry of competencies.

With these various efforts in place, both CEOs of firms and the directors of the national research laboratories will have a better map of how to build bridges between each other.

7 PROVIDING TIMELY FEEDBACK ON ORGANIZATIONAL BLOCKAGES

Step Eight in Restoring the Innovative Edge

THE ENLARGED DATA collection proposed in Chapter 6 provides managers of research organizations and business firms with a sense of context. It can help identify some external obstacles that transformational leaders such as those described in Chapters 4 and 5 may wish to address. But as is apparent from the other chapters in this book, obstacles and blockages exist at four levels. The recommended data collection by the National Science Foundation (NSF) locates obstacles at the sector (weak connections between arenas) and societal level (underfunding of manufacturing research). But the collection of data on failed evolution does not provide the kind of timely feedback that managers of research teams and the CEOs of organizations need. What they need is a new kind of evaluation model, one that measures technical progress and innovation in real time and the presence of various kinds of blockages at the organizational and team level. This requires in-depth policy studies, in fact, a new kind of evaluation research.

What is the difference between technical progress and radical innovation? In Chapter 6, measures of substantial advance for each of the arenas as well as for the sector outputs were suggested, maintaining our continual emphasis on radical innovation. But when we shift to the team level, it seems more appropriate to discuss measures of technical progress precisely so that feedback can be much more timely. There is a good reason why projects may take three to five years and why researchers frequently complain about the stability of funding: the scientific breakthrough or radical technical advance does not happen every day or even every year. Measuring the *degree of technical progress* or learning that has been achieved toward some scientific breakthrough or radical technical advance is a more realistic approach that has three advantages.

First, the new evaluation model measures incremental advances including the amount of learning that is occurring before the radical outcome is achieved. *And here is a paradox in the argument in this book, namely, that while I stress radical innovation, I recognize that it is composed of many small steps, including failures. But if learning occurs from the failures, then in the next iteration, the radical advance becomes more possible, which is why learning needs to be measured and appreciated.* This new model is in contrast to stage-gate analysis process where black-and-white goals are established rather than the more realistic gray measures that recognize incremental technical progress and even learning as more important. Stage-gate analysis is built implicitly upon the concept of productivity and concerns about efficiency in research, rather than creativity and technical progress, and results frequently in abandoning promising projects.

Second, the new evaluation model allows for much quicker feedback, within the second or third year of a project, when it is still possible to make mid-course corrections. But this is only possible if the in-depth study also isolates potential blockages or even obstacles such as the lack of cross-fertilizations or absence of enough diversity in the team.

Third, this model solves some major political problems. Budgets for research are highly vulnerable, especially in years of large deficits. They cannot be justified by scientific breakthroughs or technical advances in any one specific year. Therefore, the prudent policy maker needs to have assessments of research programs that accumulate a number of incremental advances. Indeed, the stress on major advances to justify vulnerable programs has lead to some discrediting of various claims for them that have been made, whereas an assessment of the totality of incremental advances would be much more believable.

A major, but underappreciated, obstacle in the management of the innovation process is the lack of timely information about failures to achieve technical progress in the research team or adequate levels of innovation at the organizational level. Even more important may be the lack of information about the obstacles and blockages that account for inadequate progress. To provide timely information about the lack of innovation and identifying the potential reasons for this lack requires a movement away from the traditional policy framework of evaluation, which examines the benefits once the project is completed.

A good example of the standard evaluation of science and technology is provided in the report on the Advanced Technology Program (ATP) cited in Chapter 6.[1] Indeed, this policy evaluation report is an exemplar of how to do good evaluation research, especially from an economic perspective. A number of different evaluations that were not included in Chapter 6's discussion are listed in the report. The number of publications, patents including those pending, and additional funding are good measures after the fact. The detailed case studies reported jobs created and another set of studies examined the economic benefits of some of the investments. But there are two problems with this excellent evaluation: (1) they are not timely, allowing for mid-course corrections in the management of the research projects that were funded, and (2) they do not identify blockages that might provide the basis for these mid-course corrections. In fairness to them, this was not their intent. The objective of the evaluations made of the ATP project was instead to justify continued funding of this *program* rather than to deal with the issue of quick feedback on the lack of technical progress in a funded *project* and the reasons for this lack.

What is needed is a new model of policy research for science and technology in which technical progress at the team level and innovation at the organizational level are measured in real time. Furthermore, the measures of radical innovation for each arena allow policy makers to hone in on an underperforming arena, the weak link in the idea innovation network, and do in-depth research on the organizations within that arena. Thus, the measures proposed in Chapter 6 become important for making in-depth research studies more efficient. One does not have to study the entire idea innovation network.

But this new kind of in-depth policy evaluation model also requires new measures of technical progress. It cannot rely upon the ones used in the ATP evaluation just reported if it is to provide timely feedback. Why is it so important to improve upon the widely accepted measures of papers and patents for technical progress in scientific and technological research? Particularly at the level of the research team, which is the key engine of the innovation network— where ideas are created, refined, and synthesized—measuring technical progress with papers and patents is usually too late to adjust the diversity of the team or install some mechanisms of cross-fertilization to improve learning. If team managers are to effectively manage their projects, they need more immediate feedback about technical progress at this level. By extension, the same logic applies to the boards of directors and policy makers; they have to receive more or less continuous feedback about the innovative performances of programs within the research organization or firm if the organization is of any size because it will have multiple research programs. Again, to repeat, firms may appear to be doing well on productivity but actually be falling behind in the innovation race with their competitors, which is best measured by benchmarking against them.

Developing this new kind of policy evaluation model has then two tasks. The first task in providing timely feedback is developing measures of technical progress at the team and program levels within an organization. Measures of technical progress in noneconomic research are difficult to construct, which is why universities, national research laboratories, and various funding agencies count after-the-fact papers, patents, products commercialized, and now jobs created. The first section of this chapter addresses this task of attempting to provide timely feedback on technical progress and innovation in public research programs where it is more difficult to assess progress. At the team level, an important distinction is made between measuring technical progress in technology versus science because the latter has proved more difficult, given the absence of an accepted standard of what is known in contrast to the benchmarking products. In this section, some attention is paid to the problem of measuring learning because it is one of the more neglected aspects of policy evaluation. An argument is made to measure technical progress at the program level where one can demonstrate the aggregate value of many small incremental steps. The measures of innovation at the organizational level are easier to construct and in fact are similar to those for the sector level.

The second task is to identify the obstacles and behind them the blockages within the team and the organization in which it is located. The second section focuses on looking for blockages in the research teams and organizations. Three sets of blockages at each level are discussed.

MEASURING A RESEARCH TEAM'S TECHNICAL PROGRESS AND INNOVATION

The difficulty of measuring technical progress at the level of the research team varies with the nature of the research team. Before one begins an evaluation of a project, it is necessary to be clear about the kind of project, what are the most appropriate measures for each kind of output, and what are the most appropriate time intervals for applying some measurement. In Chapter 2, I made a distinction between science, technology, and product or service characteristics.

As will become clear in the following discussion, measuring technical progress becomes much easier as one moves from the top to the bottom of Table 6.1. Advances in a technology are easier to measure than advances in science. Measuring improvements in the characteristics of the product (addition of functions, increased reliability, lower operation costs, etc.) and service (lower recidivism rate, shorter time in hospital, higher graduation rate) is still easier. Or to return to the distinction between the six arenas of research, assessing technical progress is hardest in basic research and much easier in applied research, product development, manufacturing research, quality research, and commercialization research. A radical advance as represented by the substantial improvement in a performance of a product (e.g., the movement to high-definition [HD] television is relatively easy to quantify, as we saw in the example of ferroelectric liquid crystals [FLCs] in Chapter 6, where the quality of the display image improved 600 percent).

Another useful distinction in developing measures of technical progress is the famous Aristotelian one between ideas and tools. Each research project, regardless of whether it is primarily scientific or technological, is likely to have both of these components. Indeed, in scientific research not enough attention has been paid to the improvement in measurement instruments and their capacity to measure as a necessary component for making scientific breakthroughs, which is why making instruments available at the national laboratories represents such an important asset. The reverse is true in technological research, where usually the ideas behind the technical advance are frequently not made explicit. Thinking about the simple distinction between ideas and tools allows us to recognize that research projects can learn in two ways. Thinking about these two kinds of learning in general terms, new ideas can be concepts, proof of concepts, theories, or more generally, understanding of various processes. New tools can be components or elements, instruments, hardware and software, and up to whole technical systems. The example in Chapter 6 of new software for coordination of the supply chain is a good example of a new tool.

Three sets of examples of how one can measure technical progress in different situations have been developed by the Center for Innovation and, it is hoped, provide models of what can be done as well as expose limitations and problems that we have not been able to solve. The *first set of examples* are twenty-two research projects, eleven of them in technology and eleven of them in science, funded to achieve radical breakthroughs in a high-risk research program in one of the national research laboratories funded by the Department of Energy (DOE). The proposals and the progress reports were read to determine the specific technical advances of the research objective. In this situation, one could actually measure the progress each year as incremental advances were made toward the goal. In the second example, this type of measurement was not possible and is a characteristic of the kinds of research in noneconomic sectors. We discussed our identification of the major dimensions for measuring technical progress with the researchers, and in about 95 percent of the twenty-two high-risk projects, they agreed that indeed we were correct. This is an important point because it means that individuals not trained in the specific scientific disciplines or engineering specialties can select measures of technical progress on the basis of project reports and proposals.

As a group, the eleven technology projects had a number of common themes, ones that one might have been anticipated given the evolutionary patterns discussed in Chapter 1. In other words, they provide an additional check on the ideas advanced there. So, these are the common aspects of technical progress:

- Greater reliability
- Higher efficiency
- Better precision
- More flexibility
- Increased miniaturization, or reduction in size
- Ease of use
- Speed of use

The dimension of flexibility is particularly interesting given the desire to have technologies work in a variety of situations. Likewise, the ease of use and speed of use are important considerations.

Obviously, these general criteria do not measure improvements in functionality, which is, of course, the major goal of the research. But it reflects the multiplicity of objectives in research, namely, functionality must be improved, yes, but while holding constant constraints such as size, cost, reliability, and so on. What were some of the kinds of objectives of functionality being researched? One of the technological projects that we examined was trying to improve three-dimensional (3-D) imaging technology, an area of extreme importance. Another project that emphasized the importance of measurement was the creation of new microscopes to work at the gene level, including a single molecule, high-speed fluorescence resonance energy transfer (FRET), coherent anti-Stokes Raman scattering (CARS), and so on. One of the more interesting projects relative to the problem of reducing externalities was one that was concerned with the design of the analytical converter, especially for diesel engines; the team was able to achieve an enormous increase in the amount of particulates captured. In these cases, radical innovation was measured by the percent increase in a key performance.

Many projects, however, were interested in improvements in multiple performances while holding constant constraints, as I have observed already. With multiple performances, one computes the percent increase in each performance weighted according to importance. A digital radar project, where analog was replaced with digital, was able to accomplish multiple objectives, which is generally true, as we saw in the discussion of digitals' impact on telecommunications in Chapter 1. Another project was focusing on advances in radiographic technologies with improvements in brightness, penetration power, and decrease in size of apparatus. Still a third was trying to design and fabricate a tube within a tube with very thin walls and in micro size. The importance of developing good instruments is illustrated in the following two projects: one was the creation of next-generation instruments for direct observation of chemical and physical processes in living cells, the other was an attempt to develop testing equipment to measure both vibration and shock.

What were some of the eleven science projects in which it was more difficult to develop measures of scientific advance? One project concerned the development of high-performance algorithms and software for massively parallel computer systems. Another project focused on discrete mathematics and combinations in applied mathematics. Still another was examining the interface between organic and inorganic materials at the level of nanoscience. Yet another project was working on eukaryotic organisms for biomass degradation, while still another was concerned with developing new catalyst formulations and demonstrating their utility in process aqueous-phase applications.

How does one quantify scientific advance in these disparate projects? Oddly enough, one common thread through most, if not all of them, is the idea of the amount of variance explained, a statistical measure. Implied in most of these projects was something to explain. Insofar as it could be explained and statistical techniques are one way of measuring variance, one can say there has been an increase in understanding. The extent of the increase in variance explained then became a surrogate for the radicalness of the advance in scientific understanding. A large increase in variance explained would imply a large increase in understanding. One reason why this particular approach worked well in this national laboratory was it had developed simulation techniques and therefore relied on this way of thinking as part of its technological tool kit. However, it must be admitted, we could not develop quantitative measures for all the science projects. One of our failures was a project concerned with the description of plasma fields. As much as I would like to quantify each aspect of technical progress, I must admit that not everything can be quantified.

As is obvious, some research projects do not achieve their goals immediately in part because they discovered that the initial idea or technical approach was flawed. Rather than treat this as failure, it should be recognized as learning about what does *not* work. If the reasons why the initial idea was wrong can be understood, then clearly something has been learned. By this criterion, a project is a success without technical progress relative to its objectives. Measuring learning, or attempting to, is not always easy but indicates in another way how the proposed new policy evaluation model based on an innovation perspective leads to new insights. The brief discussion of the learning process with the Wright brothers in the middle of Chapter 3 makes this point dramatically.

This list describes a number of ways in which learning can occur:

- Determining that one or more approaches will not work; specification of which ones
- Identification of problems that need to be solved; specification of which ones
- Identification of solutions to specific problems; indication of the nature of the solution
- Addition of new technologies; research strategies
- Change in the direction of the research reflecting what has been learned
- Choice of new architecture among several; identification of reasons for choice

This list may seem a little obvious, but in fact, most research papers, reports, and the like tend not to report this kind of information. Instead, the focus is on what was achieved and more typically milestones, which are either-or statements. As can be seen in this list and consistent with what has been said before, learning involves recognizing that there are problems that have to be solved as well as finding the solutions to these problems.

The *second set of examples* reflects approaches one can use to measure improvements in services. Unlike the examples previously given where real time involves determining the degree of progress after two to four years on the basis of progress reports, the nature of service projects, in many instances but not all, requires waiting to determine the effects of some service intervention such as a new cancer treatment or a school program for slum children. Unlike the science and technology programs described in the first set of examples where the measurement of technical progress can be more or less continuous, this is usually not the case in service programs. In them, the determination of the effectiveness of the intervention sometimes requires waiting a long time, such as the study of the effect of Head Start on children ages three to five on their entry into high school some eight years later. In fact, one of the more interesting issues in measuring the effectiveness of service is how long the effect lasts, or the delay in health relapse, or recidivism in prison treatment programs.

These examples originate from a large ongoing project in measuring technical progress and its benefits in medical research. The Center for Innovation is examining six morbidities: breast cancer, melanoma, colorectal cancer, Alzheimer's disease, and frailty and falls, the latter two having been combined because of the similarity of the treatments being studied. In coding the technical progress, a set of sixteen metrics are being used to detect advances and they reflect four different ways of intervening in the disease process (i.e., kinds of research: prevention, diagnosis or prognosis, treatment, and posttreatment). In each service area, it helps to distinguish kinds of research projects in a similar fashion, that is, by either stages or kinds of interventions. This stimulates ideas about new metrics that can be used.

As an illustration of research on prevention, one medical research project looked at the onset of Alzheimer's disease. As a number of patients in the early stages of this disease are also depressed, this study required three years to determine if there was a delay. A clinical trial indicated that treating the depression with a standard antidepressant drug delayed on average the onset of the more severe symptoms for about three months.[2] While this amount of delay may seem trivial, it is not for the many members of the family who are providing home care for their parents or grandparents.

A study of melanoma provides an example of the importance of medical research on the speed of diagnosis and the accuracy of prognosis, two other metrics of the sixteen. A new staining technique has been developed that is used on what is called the sentinel lymph node, that is, the one closest to the diseased mole, when a patient is first diagnosed with melanoma.[3] As is well known, the problem with this particular disease is that most patients are not diagnosed until after it has already metastasized, and then the prognosis for survival is low indeed. This staining technique diagnoses metastasis much earlier than standard observational techniques (e.g., CAT scans or physical manipulation). If the sentinel lymph node indicates metastasis, then it is removed. The early diagnosis led to a considerable increase in survival rate because of the earlier surgical intervention: 22.5 percent reduction in mortality at the end of the fourth year on average, which in medical research is an enormous gain. This large clinical study lasted eight years, the length of time being important because the usual standard in cancer research is at least five years to be sure that there is no recurrence. Another important contribution of this research study was also a considerable reduction in the number of new tumors that appeared. In other words, there was a delay in the return of the disease for some patients and a reduction in its severity. Furthermore, this same mapping technique is an example of a generic technique that can be applied to some other cancers.

Treatment programs are, of course, the more typical kind of medical technique. An interesting example of one of these again comes from the melanoma research program. Tumors from a patient with the disease are used to grow vaccines in the laboratory and then injected into the patient along with various chemicals to help stimulate the growth of antibodies. This experimental treatment for melanoma at the National Cancer Institute in the United States is one that has had an impact on a number of the metrics that we are using in our research. It considerably reduced the amount of time the patient has to spend in the hospital when receiving the treatment, it increased the survival rate from 15 percent to 50 percent, and it improved the quality of life during the treatment.[4] Under these circumstances, one adds the percent change in each metric to capture the complete treatment impact. Multiple impacts would be indicative of a major breakthrough in treatment. However, it is also important to recognize that, in this case, the breakthrough occurred after some twenty years of continued research by Dr. Steven Rosenberg and his teams in which there were many dead ends and continued learning. The sudden leap in progress would probably not have been possible without this prior effort, again stressing the importance of learning.

Several examples of posttreatments, as distinct from treatments, can also be scored with these metrics. In a clinical trial for colorectal cancer, the

provision of aspirin appears to reduce the reoccurrence of the disease, depending upon the frequency of use (daily is better) and the dosage (one-half
is better), although there is some disagreement in the literature about this
posttreatment.[5] Another example of a posttreatment and one that is extremely
interesting because it could have widespread effects in other cancer cases is
the provision of psychological counseling to women who have had breast cancer. This clinical trial, again over an eight-year time period, indicated that the
treatment group had fewer relapses, lived longer, and had a reduction in other
kinds of illnesses.[6] Again, this study concerned a medical intervention with
multiple impacts as measured by the metric system.

Admittedly, it is much easier to develop measures for clinical medical research because it is the counterpart to product development (i.e., the main
emphasis is on tools and techniques, rather than ideas or science). Developing
measures of understanding in medical research is clearly more difficult but
potentially possible. This is an important consideration because about three-
fourths of the research in each of these programs is either in basic science, the
process of the illness, or applied science, learning what kinds of tools or techniques should be developed. Although this set of metrics is for health care, the
logic can be applied to other noneconomic service systems such as education,
welfare, and so on. The issue is to perceive that it is a process with multiple
stages and that in each stage one tries to develop metrics that recognize both
the quantity of improvement and the quality of improvement.

The previous set of examples involved services. The *third set of examples*
concerns products developed by a mission agency, the National Oceanographic
and Atmospheric Administration (NOAA). As is apparent in the discussion of
Chapter 6, the national research laboratories have also been involved in cooperative relationships to develop products. Therefore, paradoxically, product development becomes another way of measuring progress in research. The difference is that the products being developed by the Center for Satellite Applications
and Research (STAR), one of the research arms of NOAA, are provided free and
are not commercialized. What are some examples?

The basic products of STAR are algorithms that translate signals from satellites into useful measures of the atmosphere and ocean, many of which are
provided to the National Weather Service. Again, as in the study of the national
research laboratory reported earlier in this chapter, we examined in this instance eight projects. They were selected to represent the range of projects across
the three major research divisions of STAR. As we examined the objectives of

these research projects, one valuable lesson we learned was that not all of their projects were concerned with developing new algorithms. Several important projects focused on improvements in the quality of the algorithms and reductions in their error rates.

Research involving calibration and validation does not stop with the determination of causes of error in the data that is collected. More fundamental is research about the methods of calibration and validation, which exist in STAR and might be called research technology, an ignored area in many standard science and technology evaluations. Examples are the new methods used to correct for errors in what is called ground truth (the actual temperature or humidity in a specific location) or the creation of new measures of ground truth, which are problems in how to validate the data. Another important issue for STAR is reconciling the information received from different satellites, and one project found a technology for assessing satellites as they passed over the polar regions. This becomes particularly important in connecting information from different satellites and across time as satellites are replaced. Another research project involved learning how to compress more satellite data, which anticipated the new generation of satellites that would provide a considerable increase in data points.

Research for the purposes of improving and/or developing new methods, new algorithms or products, and new time series are the more easily recognized examples of research within STAR. However, here too, it is useful to draw a careful distinction. Algorithms can be improved when errors are detected and corrected as well as from research on the methods of assessing error just discussed. The more fundamental and innovative improvement in a product occurs when a new data stream or variable is added to a product, in particular if it results in an improvement in its predictive ability, again the variance explained. In hurricane forecasting, a new variable improved the predictive ability by about 5 percent. Likewise, the development of a new product is an innovative activity. These examples show the value of developing a framework for measuring technical progress, leading to new insights about the kinds of research that might be accomplished with benefits for the larger organization, in this case NOAA.

What are some examples of the new kinds of products that are being developed? STAR has developed a most interesting early warning system for the bleaching of coral reefs, whose customers are largely other countries. Another new product measures the development of harmful algae in the Chesapeake

Bay in this country. Still a third detects forest fires in any part of the world. Although these products are not commercialized, they have enormous value for the world—and they are free!—which leads to the issue of whether one should measure economic and social benefits as well.

A potential defect in the new policy evaluation model is that by measuring only technical progress it ignores the economic or social benefits of a particular technical improvement. Given the influence of economic cost-benefit studies, which as we have seen are usually completed after the fact, some policy makers would like to know what these benefits could be, even though the research project may not be completed, to say nothing about a new product, technique, or service having been offered. In the Introduction, I said that one of the advantages of measuring technical progress was political, given concerns over funding. These same concerns may need a more robust justification of a research program than metrics showing technical progress. As I suggested above, if one measures potential social and economic benefits, it is better to do this for an entire research program, but again in real time, that is providing feedback relatively quickly.

Even though various technologies or products are not widely diffused, one can indeed begin to estimate the *potential* economic and social benefits of particular tools, techniques, and products. The measure of actual economic and social benefits has to make a series of assumptions about the diffusion of specific techniques or products (e.g., that the staining technique will diffuse to dermatologists who will use it or that patients with colorectal cancer will actually keep to a regime of baby aspirin daily for years). Just voicing these assumptions underlines the differences between *potential* and actual benefits. But what might be perceived as a major limitation actually is not. What it does is highlight the importance of understanding the diffusion process of technologies and the difficulties of increasing awareness about them. The dominant model in health care is a biological one based on treatment rather than on a public health model that attempts to understand what are the obstacles to diffusion and awareness and overcome them. Therefore, making the distinction should not present political problems for policy makers but instead lead to new research to solve these diffusion and awareness obstacles.

Several of the preceding examples are from a larger project that is estimating the *potential* economic and social benefits for each of the five research programs previously listed. For example, in the medical research cited earlier, an obvious social value is the number of months added to life by a particular procedure. The economic value is obtained from examining the savings in the cost of additional operations in the case of various cancer interventions, whether new prevention procedures, diagnostic techniques, treatments or posttreatments. The Center for Innovation is preparing a book on this project of measuring technical progress in medical research and its benefits. In some of the examples from STAR, we estimated the economic value of the early warning system by examining the consequences of the loss of coral reefs on the fish populations of the world, which turns out to be substantial. This estimation, of course, depends upon arguing that in the absence of information, nothing would be done to prevent their decline, and conversely, in the presence of information, action is taken. A clearer connection between the development of a new algorithm and its economic value is the early forest fire detection system that STAR developed. The sooner forest fires are detected, the quicker the response and the less the damage and, depending upon the location, the fewer lives lost.

The book being prepared on the measurement of *potential* economic and social benefits of the five research programs in the National Cancer Institute (NCI) and the National Institute on Aging (NIA) is designed to illustrate this new policy evaluation model that tries to measure technical progress in real time. In this instance, the subject will be all the clinical research studies that have been completed over the three-year period from 2006 to 2008, recognizing that some of these completed studies were started ten years earlier. The key is that adding incremental advances across research projects in a program and estimating their *potential* economic and social benefits, rather than searching for the significant breakthrough, is politically much more viable. The significant breakthrough only justifies its program and not any of the others. By comparison, the new model is more believable because it assesses all the relevant research projects. I believe that the new evaluation model would be more acceptable to Congress and the Office of Management and Budget (OMB). Therefore, I recommend that federal and state government research programs develop metrics to measure technical progress.

The measurement of innovation at the organizational level is more straightforward. Typically, it would account for the number of innovations made in a specified time period, usually three to five years. For example, for STAR, the Center for Innovation counted the innovations over a three-year period. Business firms typically use five years, and of course, the amount of time depends upon the average development time for the specific sector. Drug companies spend much longer time periods developing new drugs. Beyond counting the number of innovations, one would use exactly the same set of measures I suggested for the sector output in Chapter 6:

- The degree of radicalness on the various dimensions of the product mix
- The average speed of product development or time to market
- The competitive position of the product mix

In summary, a number of ways of providing feedback on technical progress have been presented for one major reason. Consistent with the argument that each sector is different, metrics of technical progress appropriate for one sector do not apply to another. Certain analogues do exist and a number of these have been cited, but in the final analysis, it is the specificity of the measures that is important and at all three levels: team project, research program, and organization. For example, a major distinction is made between the relative ease of developing metrics for technologies and the much more difficult task of doing the same for the sciences. A comparable difference exists between the economic sectors and noneconomic ones. But despite the apparent difficulties, it is important that metrics of technical progress be developed for public research programs in order to be able to justify continued spending on a particular line of research, especially given the current budget crisis. And this same crisis requires measuring technical progress in real time within the constraints of how long it takes to obtain reliable information. In other words, the objective of this new kind of policy evaluation research is to address the very real political problems of funding research. But it has another value as well. Another objective of these measures is to detect failures, which then raises questions as to their causes, our next discussion topic.

FINDING OBSTACLES AND BLOCKAGES IN TEAMS AND ORGANIZATIONS

In looking for obstacles in the characteristics and processes of research teams that impact on technical progress, the feedback should start with an assessment of the amount of team diversity and practices for encouraging cross-fertilization.[7] But as I have repeated several times in this book, behind the obstacle may be a considerable number of blockages. A research environment survey has been developed by Jordan (2006) that can be used to detect six different kinds of blockages: the first three at the team level and the second three at the organizational level.[8] The six sets of blockages are the following:

1. Absence of team processes associated with innovation
2. Lack of certain kinds of communication
3. Inappropriate managerial style
4. Inadequate reward structure for researchers
5. Not enough research resources
6. Absence of a long-term organizational strategy

The *first* set of blockages measured in the research environment survey has five different attributes to measure conditions that encourage creative breakthroughs within team processes independent of the diversity of the research team. For example, researchers are asked to report what percent of the time they can characterize their work as providing them with a sense of enthusiasm, time to think creatively, freedom to explore new ideas, ability to take large risks, and make decisions about their research. Most of these attributes have figured in empirical studies on research, typically relative to new product development in firms.

One important theme argues that researchers need to experiment and take risks in order to be innovative and that control stifles innovation and creativity.[9] Therefore, experimentation and taking risks provides opportunities for learning, including, as I have observed, learning that certain techniques or methods or ideas do not work. This theme also highlights one of the dilemmas that face research managers that we have already discussed, the freedom to take risks and the need for control. The case study reported in Chapter 3 provides one solution. As was suggested, the key is a balance between individual research objectives and organization strategic goals. Paradoxically, researchers on product teams who engaged in more experimental or improvisation of product design through frequent iterations, more testing, and power leadership developed products more quickly.[10]

Providing time to think and to explore new ideas are obvious issues in the way in which managers can facilitate the creative processes in the team. An important attribute of the research environment survey is that researchers are asked to report how much each of these process characteristics is present and then asked to indicate how much they think it should be present. The differences allow managers to fine-tune their practices while the survey gives them legitimacy to make changes that facilitate these processes.

The *second* set of blockages expands on the theme of cross-fertilization, including time spent on technical communication. This includes the ideas of critical thinking, exchange of technical ideas with individuals within the same discipline, exchange of ideas with those in different disciplines, technical communication within the project and external to the project. Distinctions between internal and external teams are important for the research manager because of the growing number of interorganizational collaborations of one kind or another. One review of research found that both internal and external communication were associated with successful launches of new products.[11] The case of IBM illustrates this idea of strategic alliances or external communication. Again, on each of these dimensions, the researchers who are given the survey are allowed to express how much each of these attributes should be present to provide the kind of feedback that managers need to correct the amount of cross-fertilization.

The *third* set of blockages focuses on the very important role of managers, especially those who are the immediate supervisors of the various research teams. The research environment survey allows researchers to assess the following dimensions of intermediate managers: integrity, technical value added, overall value added, planning and execution of projects, and appropriateness of the criteria used to evaluate technical progress. Most people would agree that these are important characteristics, and they do not need much comment.

The next three sets of attributes shift to the organizational level and various practices that may represent blockages to scientific breakthroughs and radical technical advances. Generally, the reward structure, the *fourth* set of blockages in the list, reflects an organizational policy rather than anything that can be controlled by the manager except in how they apply criteria for measuring technical progress, which I just mentioned. Too many people think that the key reward is salaries. This may be true for the accountants and stockbrokers that have made fortunes recently, but it is generally not true for researchers, who are motivated much more by solving problems, the intrinsic reward of their work. Both intrinsic rewards, such as recognition for merit and the respect that researchers receive, and extrinsic rewards, including salaries and fringe benefits, opportunities for career advancement, and for professional development, are measured. Analyses of the attitudes of researchers in various public research laboratories where the research environment survey has been used indicate that respect and recognition of merit are more important than salaries and fringe benefits.

The *fifth* set of potential blockages refers to the quantity and quality of organizational resources for conducting research. Achieving radical innovation is not only a question of solving problems; it is also a resource question. The list of resources includes most items that anyone would expect: resources for exploring new ideas, equipment for research, the physical environment, stability of funding, quality of the technical staff, and staff mix well suited for the execution of the researcher's skills. Of these, the stability of funding is a major issue even in public research laboratories, and contrary to the general image of the civil servant qua bureaucrat who works for the government. Researchers are expected to raise research funds in many areas of the federal government through proposals. What is not well appreciated within the federal government are the long-term consequences of the lack of stable funding; it seriously reduces the innovativeness of research for a considerable time afterward. A study of high-tech biotech firms reported in Chapter 2 found that firms with a serious funding disruption tended to be reluctant to spend their funds and thus explore new ideas.[12] Another consequence is that top managers who were previously scientists might be replaced by business managers, as happened in one firm; and like the accountants at GM, they instituted tight financial controls that had the same consequence. Many government researchers feel that they are torn between trying to continuously raise money and having the time to do research. This is a potentially large blockage that needs some careful research.

In addition to these resource questions, there are also three questions about support, including the availability of top computing and library services; the cost-effectiveness of procurement, security, and safety; and finally, another critical one that is a frequent obstacle, the competitiveness of the overhead rates. The rates have become quite large in some of the national laboratories, thereby affecting the ability to attract external funding.

Finally, in the search for potential blockages, the *sixth* set provides feedback about the researchers' perceptions regarding the overall strategy of the research organization or firm and how much it is oriented toward the future. Under this theme, there are seven attributes: vision; strategies for carrying out the vision; the process for identifying new programmatic opportunities; investment in new programmatic opportunities; the depth of key research and technical competencies; a research portfolio that includes basic, applied, and development research; and allocation of research funds across areas and ideas.

Consistent with the arguments about evolution, the expansion of the knowledge base, and the continual increase in demands for higher standards, top management has to identify new programmatic opportunities and investments in them. To do this also requires the hiring and the development of collaborations to have the right set of competencies. Lacking movement into new programmatic opportunities and seeking the right kind of diversity for these opportunities can prevent research teams from creating scientific breakthroughs or technological advances. Examples of how important this is were provided in Chapters 4 and 5.

What is the value of this research environment survey in providing feedback about these various kinds of blockages? It was used in the STAR division of NOAA in 2005, and a series of recommendations were made by the Center for Innovation, especially to increase the rate of communication. The head of the division, Al Powell, Jr., took the recommendations and, as a consequence, in the next wave of interviews in 2007, there was considerable improvement in the rate of communication as well as general satisfaction with the research environment. But more critically, the rate of product innovations had jumped fivefold.[13]

In summary, many practices or routines at either the team or the organizational level can be blockages that prevent the development of radical product and process innovations. The strength of the research environment survey is that it contains fifteen potential blockages that can prevent teams from achieving radical innovations. And of these fifteen, the researchers are allowed to express their opinions about ten of them. This feedback, when combined with an assessment of the kind of diverse team that has been selected and its appropriateness for the strategic goal, its kind of idea innovation network, and the presence of mechanisms for cross-fertilization, provides a powerful set of tools for detecting potential causes for the absence of radical innovation at the team level.

At the level of the organization, the context for the team, another twenty-three potential causes of a lack of radical innovation exist in the research environment survey. In addition, the evaluator can use benchmarking techniques with the various strategies suggested in Chapter 1. When combined with the characteristics of transformational organizations, the evaluator can provide useful feedback to the managers of the research organizations and firms.

EVOLUTION AND THE NEED FOR TIMELY FEEDBACK

Again, this chapter provides a number of insights about how to make both research teams and organization more open to ideas in their external environment. The starting point is to recognize that we do have something to learn, not necessarily an easy position for many researchers who, by definition, pride themselves on their expertise. But all of us have limits to our knowledge and we need to expand them. Therefore, the key to starting this learning process is to measure the extent of technical progress in the research team and program within the organization.

The evolution of science and technology presents the United States with a number of opportunities to move ahead of the innovation curve, but it comes with a price—the need for constant feedback on the measurement of technical progress and research and the presence of obstacles and blockages that can explain lack of progress. The pace of changes precipitated by the growth in RDT expenditures and the globalization of research requires continuous assessment as the speed in the creation of radically new products and processes accelerate. Indeed, one of the striking differences of an evolutionary approach to scientific and technology policy is the recognition of change and flux. This has made both complexity theory and chaos theory popular, but as this book has tried to demonstrate, rather than chaos, predictable patterns in evolution exist and evolutionary failures can be anticipated. With the price of constant feedback comes the benefit of learning how to manage the innovation processes better at multiple levels for many managers.

Just as we need an evolutionary policy model, we also need a new model for evaluating science and technology, one that focuses on the extent of technical progress in real time or the lack of it, as well as the various blockages at the team and organizational level that might explain this. But there are problems with this new policy model for evaluating technical progress in research. In some service projects, the consequences of the service intervention only become apparent after some time, which means that intervention in real time is not really possible. In these cases, all that could be done is a retrospective study to try and estimate how the research project might have been able to develop a more effective intervention. Science projects present a special challenge because in some sciences, such as botany and zoology, description rather than analysis is of course acceptable. Under these circumstances, it becomes difficult to describe technical progress.

Ideally over time, special metric systems can be developed for each of the service areas and for specific specialties within science. The Center for Innovation has developed one for clinical research in medicine that can provide some general guidelines, but the reality remains that each scientific, technological, product, and service sector is different and needs its own kind of metrics for assessing technical progress at the team and program level. The federal and state governments need to focus more attention on developing measures of technical progress in research in noneconomic sectors other than health, such as education research and welfare research. Even more important is the establishment of measures for the new kinds of global problems, such as research on terrorism, global warming, dependency upon oil, and substitution of new materials for metals and woods that are scarce.

More rapid assessments of technical progress are important given the desire of the OMB to eliminate programs that are not working. Justifying various research programs has thus become a major political issue, which is why we need my timely feedback. Assessing technical progress in research is therefore useful as policy information, even with the limitations noted previously. But before programs are discontinued, it would be wise also to consider the amount of learning that has occurred, which is why I have suggested some ways in which this could be measured.

What will be learned from the constant feedback, which is the price of restoring the innovative edge? An additional bonus is the building of the science of science and innovation policy. The connection between technical progress and various kinds of blockages should provide a new understanding about the management of innovation. In addition, connecting this information to the technical progress in an arena, as well as to the more expanded data set that I recommend be collected by NSF in Chapter 6, would provide a considerable amount of learning that should lead to new policies that can accelerate the speed within which scientific breakthroughs are translated into radical product and process innovations. This is a particularly important area for further research, given the policy objective of moving more scientific discoveries into profitable industrial innovations that protect employment.

EPILOGUE
A New Socioeconomic Paradigm

It is only by building a new foundation that we will once again harness that incredible generative capacity of the American people. All it takes are the policies to tap that potential—to ignite that spark of creativity and ingenuity—which has always been at the heart of who we are and how we succeed.

President Obama,
August 1, 2009, in a radio address to the nation

RELATIVE TO THESE words by President Obama, the two objectives of this book have been (1) to provide a new foundation with a socioeconomic paradigm, and (2) to provide a new set of policies that can tap that creative potential. The new foundation starts with a shift away from emphasizing productivity as the goal for economic growth toward innovation in products and processes, and not just any innovation but radical ones. The new socioeconomic paradigm that provides the foundation is built around the concepts of evolution and failed evolution. The latter idea leads naturally to a discussion of what the new policies to unleash the "sparks of creativity and ingenuity" are.

The new socioeconomic paradigm has four essential ingredients that can be usefully summarized in this epilogue:

1. Predictions about the processes of social change
2. Principles of organization to create radical innovations
3. Practices of cooperation
4. Perspectives on economic growth

These four *P*s form a review of how this new foundation can be constructed.

But perhaps the most important question for any policy maker is to explain failure and to know what to do about it. A new policy model consistent with the socioeconomic paradigm has been proposed, and it has the following characteristics:

1. Collecting new data at the *societal* and *sector* levels to document failed evolution or valleys of death and various obstacles
2. Supporting timely feedback at the *team* and *organizational* levels to discern blockages
3. Adopting new manufacturing policies to create more employment

These form the basis of the new policies for unleashing the creative potential in the country.

A NEW FOUNDATION

The *first and most important way* to evaluate the usefulness of a paradigm is to determine whether it can predict change and therefore explain the need for a new foundation. The conceptual heart of this new paradigm is the ideas of evolution and of failed evolution. Evolution in how radical innovations are produced creates a complex process that is itself *evolving across time.* The initial trigger mechanism in each sector is the growth in RDT (basic research, applied research, and product development) expenditures both nationally and globally. The mechanism that explains the evolutionary processes is the need for greater focus, which creates new occupational specializations in research and technology, the emergence of new research organizations, and the differentiation of one or more of the six research arenas. But other changes have to occur for the pace of radical innovation to last: the creation of different kinds of networks, increases in the strength of the connection, and more network coordination.

A key argument in this book has been that this evolutionary model unfolds at a different pace in each sector, in part because of the differences in the amount of RDT expenditures and in part because of the disparate ways in which these sectors are organized, as described in Chapter 5. *But a nontrivial implication of this principle is that internally the sectors of the United States are becoming more and more different!* Add to this the globalization of research and the patterns of success and failure in the world marketplace, and some sectors are growing while many others are declining, creating even more differences. The great advantage of this paradigm is that it takes what many perceive as chaos, to use a popular term in academic circles these days, and turns it into something that is highly predictable, understandable, and usable for informing policy.

The other major evolutionary force originates in the rising levels of education. The movement from an average secondary education to mass university

educations has led to changes in tastes toward more technological sophistication, better design, higher quality, and also a willingness to pay for a reduction
in the risks to the environment, but most critically, demands for customized
products and services to fit a highly specialized set of needs for each individual. The need for the customization of products and services not only within
this country but for those who live in other countries is one of the biggest challenges facing the United States because this idea is so contrary to the way in
which this country industrialized, which was based on the principle of "one
size fits all." One advantage of this paradigm is that it makes this challenge apparent and therefore tractable.

But while the evolutionary processes are increasing the differences between
sectors in society and between societies, there is paradoxically a direction
to these change processes: the relentless movement toward greater complexity
both as a cause and a consequence of the evolution in how radical innovations
are created. The term *complexity* is a much-overused word and frequently
abused as well because it is not well defined. In this book, there have been two
concrete definitions, one is the way in which the processes of knowledge production are changing, as defined in Chapter 1, and the other is the way in which
the research problems that have to be solved are themselves altering, which has
been mentioned at various places. The implications of this growing difficulty
in the research problems are that business leaders and policy makers must
keep creating more complex responses as evolution in science and technology
unfolds.

At various points, the following concrete dimensions of difficulty of the
problem have been touched upon:

- The cost and risk (meaning choice of wrong solution)
- The variety of arenas that have to develop radical solutions for radical
  product innovation
- The need to eliminate multiple kinds of risks to the environment and
  the individual
- The requirement of global solutions that can be customized to specific
  regional and national differences

Over time, more and more scientific problems are perceived to be complicated both because of external influences that must be taken into account and
internal processes that must be modeled more or less well to improve the quality of the prediction. In science there is a steady movement toward viewing

problems as systemic, with multiple levels and alternative internal processes, exactly the way in which this book is constructed. In other words, research problems are growing in scope. At various points, we have observed how technological problems have steadily become more difficult, in part pushing competitors to join together. The 300 mm silicon wafer is not the first, nor will it be the last, technological problem that creates a grand international alliance. Perhaps less appreciated is the fact that it is applicable to many of the high-tech sectors, whether airplanes or nuclear energy, automobiles or high-speed trains. Another way in which to describe how problems have become more complex is the recognition of global problems such as terrorism, genocide, global warming, dependency upon oil, scarcity of basic raw materials and foods, pandemics, and so on, none of which have easy solutions.

The central thrust of this book is that just basic and applied research with a little product development thrown in do not suffice. The idea innovation network offers a framework to connect the various arenas of research; eliminate blockages, silos, stovepipes, and valleys of death; and learn from failures in evolution in order to compete successfully in a global economy. We need to continuously focus on manufacturing research, quality research, and in particular, eliminating externalities. I have placed considerable stress on the importance of manufacturing research and quality research to give American firms a better edge in the market. Less stressed is the implication that improvements in these areas will likely require radical innovations, and certainly the suggestion of a third industrial divide is that. More concretely, as problems become more complex, they necessitate radical solutions in each of these six arenas of the idea innovation network.

Equally complex is designing products so that they reduce the external costs to the environment. In particular, the question of how to develop radical manufacturing processes that use little or no energy is a problem that will not go away. This is most obvious in the development of the electric car and other energy-efficient solutions to transportation, but it applies to any manufactured product.

Another important criterion for testing whether a new paradigm is useful is the number of insights it has relative to existing literatures and paradigms within the various social science literatures. The end of each chapter has considered how the concepts of evolution and failed evolution as well as the subject matter add to these various perspectives. For the national systems of innovation, this paradigm highlights the necessity of shifting the focus to sectors and with this perspective how much easier it is to connect the societal level with the organizational and even the team levels. Chapter 1 provides an evolutionary perspective to the strategy literature within management as well as a new policy perspective.

In turn, the strategy literature is connected to the atom of innovation, the construction of diverse research teams. Three different literatures relative to this were synthesized and then connected to the knowledge paradigm within organizations. Managerial practices for cross-fertilization at multiple levels help explicate how organizations can learn, a theme returned to later in the discussion of cooperation. The concept of transformational team leadership is another important set of insights, especially how it has been defined and illustrated in several cases and historically.

The more complex responses that business managers and policy makers have to provide to handle the growing difficulty of problems represent *the second way* in which the new paradigm provides a foundation. Given the movement toward more difficult problems, what should be the responses of both the private and the public sectors to handle this difficulty? The answer is to increase the complexity of the organization of the processes of innovation.

Which leads to the next important criterion for evaluating this paradigm, and that is its practices for managing the innovation processes or how to produce radical innovation, the stimulation of "the creativity and ingenuity of the American people." These practices are a useful way of summarizing another part of the book:

- Increase the diversity of the research teams
- Add more practices to ensure that cross-fertilization occurs within and between teams
- Ensure that there is overlap in the specific specialties that are at the boundaries between differentiated arenas
- Integrate the diverse research teams within the organization
- Appoint a transformational leadership team
- Change the context of the organization
- Increase network coordination of the idea innovation network
- Improve public and private sector cooperation
- Collect more data on sector performances and their evolution

These nine principles provide guidelines for what a more complex response would be to the increasing difficulty of research problems as well as the description of how to create radical product and process innovations. Each of them summarizes a number of recommendations.

The two most unusual aspects of this long list are the *transformational leadership teams* and the improved *public-private sector cooperation,* because they are the opposite of two standard American myths—that the single leader will save the company, and that private enterprise can handle all competitive problems by itself. The best proof that these myths are no longer relevant is the recent economic history of the United States. The rise of many of the new companies that have become household words, such as Hewlett-Packard, Microsoft, and more recently, Google and Facebook, as well as other dot-com companies, began in garages; these were not started by a single individual tinkering in that garage but by several individuals, even if one face was more visible than the others. But we would not have had these companies in the first place if it had not been for the large amount of government support in the research on computers and then later in the creation of the Internet.

As I have argued throughout the book, a transformational leadership team is a group of leaders that have a vision larger than one company and recognize that problems have to be solved in the larger environment. The case studies of the Pasteur Institute and of SEMATECH demonstrate how important external problem solving is. Again, we would not have had SEMATECH without the support of the Department of Defense and its concerns about losing leadership in semiconductors. Interestingly enough, although not emphasized in the study of the Pasteur Institute, it too needed the help of the French government for financing and especially in the last half-century. So one evolutionary response should be the movement from single leadership to team leadership. An important issue is the composition of the leadership team. An ideal combination for such a team at a firm or a research laboratory would be a scientist or engineer, a business manager, and an expert on interorganizational relationships. This leadership team would deal with the three kinds of problems that have to be managed: innovation, productivity, and the idea innovation network, or connectedness to the external environment.

But this list leaves a number of questions that have not been explicitly addressed because they vary so much from one sector to another and even within niches of sectors: What are the specifics, relative to disciplinary specialties, leadership teams, and kinds of network coordination? The answers to these questions may require the creation of new kinds of expertise and new areas of research, which is one reason I recommended the development of a national registry of competencies within each of the arenas within each of the sectors.

However, regardless of unanswered specific questions, the general framework argues that evolution is a constant movement toward more complexity, and the best response to that movement is an increase in complexity. Without this increase and the right kind, then there is failed evolution, which then becomes a policy issue. The stimulus of more difficulty in the problem leading to more complexity in the process of producing innovations is obviously a never-ending circle.

Each of these responses has implied in it another kind of modification in our way of thinking, the movement away from the principle of competition to one of *cooperation*, a *third way* of establishing a new foundation. The big lesson of the data reported in the Introduction is that we already have a lot of competition and we are not doing very well. What we need is cooperation to fight against this competition and at multiple levels, including cooperation between the public and the private sectors. I hope that the reasons why we need cooperation have become well established across these various chapters.

The theme that has been repeated in the last five chapters is that to create radical innovation, one has to learn:

$$\begin{array}{ccc} \text{Diversity of} & \text{Cross-Fertilization} & \text{Innovation or} \\ \text{Paradigms} & + \quad \text{or Learning} & = \quad \text{New Knowledge} \end{array}$$

This cross-fertilization only happens with cooperation. Each chapter has described different ways of encouraging learning in teams, between teams, between organizations in the idea innovation network and supply chains, and last, but most importantly, between the public and the private sectors. A subtheme about learning has been the need to identify obstacles and blockages and particularly underfunding of manufacturing research and quality research, absence of important skill sets, and the special kinds of blockages that exist in organizations. All of these insights flesh out considerably how to really be open to the new ideas that exist in the environment.

To be effective in discovering scientific breakthroughs or technical advances, research members of the teams have to cooperate. In Chapter 3, I noted some of the blockages that militate against cooperation. And while diverse teams have communication problems, when these are solved, there is a multiplier effect from the pooling of their expertise. Individuals making decisions in their own self-interest and managers concerned particularly about maximizing profit and, more recently, about stock prices hinder cooperation.

Cooperation between the members of a transformational team is not necessarily self-evident, as again demonstrated in the case history of SEMATECH. For organizations, the biggest hurdle is learning how to cooperate with other organizations in the idea innovation network. Chapter 5 reported a number of failures in this regard as well as some success stories. One obvious key for creating more cooperation is a commitment to a larger vision than the personal success of the team member or leader or organization. The vision is that all should benefit.

In the study of the telecommunications and the biotech and pharmaceutical sectors in Europe, a large number of different ways of cooperating or improving network connectedness were observed: joint ventures, user groups, product teams, patent pools, collective trademarks, technology clusters, partnerships, alliances, and even virtual firms. We need to determine which of these kinds of network arrangements work well for a particular connection but most importantly how to improve it—that is the evolutionary challenge. How to connect several universities to solve some difficult problem in basic research is not necessarily the correct network to handle the supply chain of a complex product such as the airplane. Again, the principles of cooperation have to be modified for each sector and even between the specific arenas within a sector.

The paradox is that the more complicated arrangements of team and network decision making are quicker in developing creative solutions, scientific breakthroughs, and technological advances because of the greater focus achieved through the evolutionary processes, provided of course that the different arenas are tightly connected.

The most complex issues are in the cooperation between the public and the private sectors. As I have already noted, there is much more cooperation than one would assume, and various instances of this have been cited in Chapter 6. However, this still leaves unanswered the other issues that need to be addressed. What are some of the areas where cooperation is needed?

One of the key problems facing the United States in competition with many of its major trading partners (Japan, Germany, South Korea, Taiwan, China) is the poor level of technical education at both the secondary and tertiary levels in the United States. A number of historical reasons explain this absence of robust technical education in a number of the sectors, both low- and high-tech. But one of its consequences has been a lack of attention to building sophisticated manufacturing systems capable of considerable flexibility and quality. We need to rebuild this kind of education system, and especially for the kinds of manufacturing of the future if the society decides to invest in developing the third industrial divide in some sectors. Its rebuilding requires the development of networks between training centers for technical specialists and places where they can be offered apprenticeships as in the German model. In particular, community colleges are presently moving in this direction with their associate degrees, frequently in technical areas. But much more could be done, and especially outside the areas of information technology and biotechnology. This is not a new idea, but it is one that will only work with a partnership between the private and public sectors, especially at the local level.

Increased coordination of RDT expenditures is another arena in which a partnership is necessary. I have stressed the importance of transferring the focus from funding to encouraging more cooperation at the research team level, in particular in the arenas of manufacturing and quality research. I have asserted that a thrust in these areas needs examination because of the policy discussions about the valley of death. But to understand the depth of the problem, we need an elaborate data collection effort, first in the high-tech economic sectors and in those noneconomic sectors that are societal priorities, in order to have a better grasp of how much cooperation exists and whether in some sectors more is needed. This is a proper role for the federal government. Then the discussion of how to stimulate cooperation between the public and private sectors could commence.

The *fourth and final way* of evaluating this paradigm of social change is by what it has to say about economic growth. In some respects, here resides perhaps the most critical part of establishing a new foundation, because it deals directly with the problem of expanding employment by creating new jobs. The socioeconomic paradigm suggests three ways of promoting growth. The first two focus on how to rethink our measures of productivity. This paradigm, as has already been noted, even changes how productivity is defined. Rather than it simply being the cost per unit, it should be the cost per quality of unit, where quality includes the following criteria:

- Reducing defects in the product during the manufacturing process
- Lowering the repair costs during the use of the product after sale
- Extending product life
- Lowering the operating cost during the product life
- Reducing hidden costs to the environment or risks to the individual

The last criterion on the list covers a number of issues that might be separated. It includes the reduction in the use of scarce materials, less dependency upon oil, lower energy costs in general, less pollution both in manufacturing and in use, and so on. As was argued in Chapter 1, if American manufacturers emphasize enough of these measures of quality, there is the possibility of regaining some of the employment that has been lost by offering much higher productivity per quality of unit.

An even bigger way of redefining productivity is the cost across a disparate set of units. This is the issue in customization: being able to provide different configurations of product characteristics to the consumer. Development of this ability is one of the ways in which to handle the competition from Asian countries with its emphasis on large-scale mass production. If we can design our manufacturing systems so that they are highly customized and at the same time reduce costs, then we have a competitive edge. This was the great promise of flexible manufacturing, which we failed to realize. The idea of the third industrial divide, introduced in Chapter 1, offers this same promise, *if* we can both invent the new kind of technology and master it once invented. It is the latter that we failed to do with flexible manufacturing.

The reason to emphasize the *way in which productivity is defined* is to ensure that we have a clear strategic vision: reducing the cost per quality unit and reducing the cost across a disparate range of units. Hopefully, with these clear strategic objectives, then corporate boards would not reward their CEOs simply for lowering costs but only for making substantial improvements in this expanded sense of productivity. To obtain large enough increases in quality per unit and in customization requires firms to emphasize radical process innovations.

The third way in which this new paradigm contains a vision of economic growth is the creation of radically new products linked to new market niches and even new market sectors. Provided we control both the idea innovation network and the supply chain, then many new jobs can be created as we export the products to other countries. Again, in many instances we lost control of the supply chain after a significant technical advance. Here is where public and private sector cooperation of networks would be particularly useful, to prevent the movement of jobs overseas by upgrading the skills of the workers making the various components.

The entire book has focused on how to create these radically new products and processes precisely so that America can again experience economic growth and a rising standard of living. Managing the innovation process to achieve this goal is not easy, as I assume has become quite clear from the previous seven chapters.

But to maintain a coherent strategic vision about the importance of increasing economic growth via radical innovations, we should require firms to report the number of new products separate from their measures of productivity. To some extent, firms are doing this in their annual reports. I am concerned that this be made more explicit. And consistent with what was said earlier, firms should report three separate productivity figures: cost per unit, cost per quality of unit, and cost per range of product characteristics. Only by keeping these various measures separate will corporate American have a clear vision of how to achieve economic growth in a world of global competition. And CEOs will be rewarded more appropriately in terms of how they are aligning the future of their firm with the processes of evolution.

The patterns of growth in the socioeconomic paradigm of innovation are different from the competitive model built on competition between firms over price. Rather than one firm becoming dominant because it is able to mass produce products at the lowest cost, as China has been doing in many product areas, growth occurs because new sectors and niches within them are created, thereby constantly expanding opportunities for other firms to specialize in various niches created by the differentiation of tastes and technologies.

Space does not permit an elaboration, but implied in these eight chapters is an endogenous theory of economic growth. Increasing diversity and the amount of learning at each of the four levels (team, organization, sector, and society) results in generating innovation, which is endogenous economic growth. It is not growth based on productivity, except in the cases of radical process innovations, but growth based on the creation of new markets and niches with them. This line of reasoning does not see the inevitable convergence of all economies, which is the standard prediction from neoclassical economics.

In summary, this new *socioeconomic paradigm of change* has a clear understanding of how science and technology are changing. In this process of evolution, how radical product and process innovations can be generated has been specified. One key element in this process is the need for cooperation at multiple levels, including public-private cooperation. Finally, in this paradigm is a vision of new pathways for generating economic growth that leads to more jobs. But what if this evolutionary process does not follow automatically? Then failed evolution has occurred, which leads naturally into the discussion of a new policy model, our next topic.

A NEW POLICY MODEL

One of the great advantages of this new socioeconomic paradigm is that it allows policy makers to predict failed evolution and then search for its causes in the various obstacles that form the focus of the chapters in this book. This perspective is a dramatic departure from the present policy model as exemplified in N. R. Augustine's *Rising Above the Gathering Storm* (2005), which provided recommendations about more expenditures on RDT and more training of scientists. The more complex model of the idea innovation network suggests that there are eight obstacles that could potentially explain why there is an innovation crisis. Another major advantage over the current policy model is the expansion in the definition of RDT (basic, applied, and product development) to include manufacturing research, quality research, and commercialization research.

Still a third major advantage is that, rather than one general policy for all high-tech sectors, this paradigm argues that one needs to construct different policies for separate sectors, or least those sectors that are somewhat similar, as described in the typology at the end of Chapter 5. A number of reasons why the sectors are different have been provided, and presumably, given the large number of sectors that are included in this book, this point has been well established. In addition to the various arguments already made, there is one final one, namely, that the obstacles or explanations for failed evolution are likely to be different in each sector.

To realize the potential of this new policy model requires changes in the data collected by the National Science Foundation (NSF) as well as new kinds of policy research studies. As was observed in Chapter 6, in order to understand how far the processes of evolution have unfolded within each sector, we need to have data on the following:

- Which arenas have been differentiated
- How many organizations are in each arena
- What are the links between each arena
- How strong are these links

This data would provide a general context for the research organizations and firms within each sector, especially if it were possible to obtain comparative data on critical sectors in other countries. With this framework, the NSF could then report the three Cs of national policy in the six arenas of each sector: capital expenditures on research, competencies of research, and the coordination of the network. The amount of data to collect would appear to be considerable, but perhaps refinements in the data that they presently collect would reduce the magnitude of this effort.

Given some context of the extent of failed evolution, we need new policy studies that would focus on the obstacles and the blockages that explain the failures in evolution. As has been noted on several occasions, behind each obstacle are a number of different kinds of blockages. These detailed studies should focus in particular on the blockages within teams and organizations. A key element in helping to identify them is, of course, measuring the extent of technical progress or innovation, which is one reason why it was emphasized in Chapter 7.

Consistent with the idea of the need for new kinds of policy studies, I have proposed a new kind of evaluation study that attempts, as much as it is possible, to measure the extent of innovation quickly so that changes in the management of the innovation process are possible. This measurement is more difficult to complete in the service areas where one needs to wait a certain time period to determine what actually has been the consequence of a particular intervention. Nevertheless, the sooner that timely feedback can be given, the more possible it is to change the situation. Another useful feature of this new kind of feedback is that it focuses not only on measuring the extent of the innovation but on potential blockages within the team and the organization. It is this kind of feedback that is most useful for managers because it provides an analysis of how to improve the situation.

A special goal of detailed policy studies should be to find new manufacturing policies for each sector. Radical product and process innovations in the manufacturing sectors hold the greatest promise for continued economic growth and the expansion of employment. One useful policy would be to develop extension agents and technology centers for each of the low- and medium-tech sectors under the tutelage of the newly renamed Department of Industry and Commerce. By concentrating on creating new kinds of manufacturing processes—the third industrial divide—that reduce the cost per unit of quality and cost per range of differentiated products, we can potentially protect employment from continuing to migrate overseas. The recent success of Germany based on its exports of high-quality machines is an example of what can be achieved.

At the other extreme, in some of the high-tech sectors where we have fallen technologically behind—nuclear energy, high-speed trains, optical electronics—we should form alliances with the technological leaders to build the next generation. Our large market and scientific skills make our participation in such an alliance attractive for the other country. But, of course, a requirement would be that a certain proportion of the products be manufactured in this country.

In still other areas, we have all the scientific breakthroughs but are not realizing the potential of our manufacturing for various reasons. One is the lack of close cooperation between the public and private sectors to encourage better communication. Another is the lack of focus on the importance of continually increasing the technological sophistication of various components in the supply chain. This is particularly the case in both computers and information technologies and the life sciences where we have surprisingly large trade deficits. Special studies should try to locate the major obstacles. But one also suspects that it is precisely in these areas where we need to concentrate on manufacturing and quality research so that we can obtain the benefit of our investments in science and technology. This might require a better alignment between the national laboratories and various firms. If a national registry of competencies were developed, it might help in this matching process. It also might identify clearly where we are missing important skill sets and where we need to construct the appropriate kinds of training programs to create them.

In summary, the paradigm of socioeconomic change provides the basis for a new kind of policy formulation, one based on the principle of failed evolution, and new kinds of policy studies to identify obstacles and blockages that explain the failure. Focusing on the processes of innovation and managing them better, rather than just focusing on issues of investments, has indicated the need for a new policy model for evaluation research.

To return to President Obama's radio address, the socioeconomic paradigm of change builds a new foundation and one that is actionable. It explains why there is an innovation crisis and what can be done to restore the innovative edge and thus allow Americans more opportunities to be creative. It is not a simple or easy program to implement. The identification of the eight obstacles and a number of the blockages behind the obstacles indicates how complex is the task of restoring the innovative edge. But while this book has tried to deconstruct the complexity of the problem so that it can be more easily grasped, not all blockages have been identified and a number of other subjects that could have been addressed have not been. However, my hope is that this book does provide a framework on which others can build and, in the process, improve upon it.

REFERENCE MATTER

NOTES

Introduction

1. NSF, 2010: chap. 6.

2. NSF, 2010: chap. 6.

3. Associated Press, "Boeing 787 Takes Flight After Two Year Delay," December 15, 2009. What is especially interesting about this particular story is that Boeing moved to a network arrangement along its supply chain, which is discussed in Chapter 5, but has had troubles in learning how to coordinate it.

4. NSF, 2008: chap. 6.

5. Block and Keller, 2009: app. 1. I have computed the averages for the decades, subtracting the number of foreign awards in the process.

6. Block and Keller, 2009: 471.

7. Kao, 2007: 38.

8. Augustine, 2005: 219–220. An update was published as this book was going to press. See Committee Members of the 2005 Rising Above the Gathering Storm Report, 2010, which says the situation has worsened.

9. Ashok Bardhan and Cynthia Kroll, in Augustine, 2005: 211.

10. Kao, 2007: 66. In fact, IBM has largely ceased to be an American company, with 71 percent of its employees located outside the United States (Tassey, 2009: 32).

11. Jeffrey Immelt, interview by Charlie Rose, Public Broadcasting System, June 25, 2009. Furthermore, GE is building new manufacturing plants in the United States; see Chapter 6.

12. Ashlee Vance, "Acer's Everywhere: How Did That Happen?" *New York Times*, Business Section, June 28, 2009. The reasons for this success are given in Chapter 5, in the discussion of the Silicon Valley that they created.

13. Ashlee Vance, "Hewlett-Packard Fights Back with Printers Built for a Digital World," *International Herald Tribune*, Business Section, June 7, 2010.

14. Natasha Singer, "That Pill You Took: It May Well Be Theirs," *New York Times*, Business Section, May 9, 2010.

15. *Innovate America*, Council on Competitiveness, December 2004, p. 32.

16. Kao, 2007; Estrin, 2008.

17. Connie Guglielmo, "Apple Raises Heat on Android Rivals," *International Herald Tribune*, July 9, 2010.

18. NSF, 2010: chap. 6.

19. Alan Tonelson and Kevin Kearns, "Trading Away Productivity," *New York Times*, March 6, 2010.

20. Tassey, 2009.

21. Christensen, 1997; Utterback, 1994; and as is obvious, this degree of radicalness is a dimension. There are other dimensions besides the degree of radicalness; see Chapter 2. Other ways of describing radical innovation than these four dimenions are changes in the architecture of the product or services; see Henderson and Clark, 1990. Another is the idea of the hypercube; see Afuah and Bahram, 1995, which emphasizes the consumers and supplers as well.

22. Hage and Hollingsworth, 2000; Walton, 1987.

23. Zammuto and O'Connor, 1992; Piore and Sabel, 1984; Pine, 1993.

24. This last dimension is normally not considered, yet increasingily it is very important.

25. *Science and Engineering Indicators* (NSF, 2010) does not report this kind of data, nor are these ideas mentioned in any of the recent articles or books on innovation; see Arnold, 2004; Kao, 2007; Estrin, 2008.

26. Economists call these *hidden externalities*. Obviously, the oil spill in the Gulf of Mexico has made everyone aware of this problem.

27. Pine, 1993.

28. Guerrieri and Tylecote, 1998; Malerba and Orsenigo, 1993; Malerba, 2005; Pavitt, 1984. These articles approach this issue from different vantage points: one argues for the importance of culture and history, another proposes that sectors vary because of their technology, and the last author suggests a typology of innovation.

29. Hage and Hollingsworth, 2000; Tassey, 2009, calls these complex technical systems.

30. Anderson and Tushman, 1990; Utterback, 1994; these reference works explore mostly old dominant designs. We really need much more recent research on the pace of technological change in dominant designs outside of semiconductors, which as everyone knows is still dominated by Moore's law of technological change. In this regard, Christensen's study (1997) of hard disk drives provides a better insight of the pace of technological change in the high-tech area.

31. Lafley and Charan, 2008.

32. For example, an Italian baker had the solution for one problem; see Lafley and Charan, 2008: 136.

33. Hage and Meeus, 2006; furthermore, there are five disciplines represented as well. The field is fragmented, which is one reason why this book attempts to tap into a number of different literatures.

34. Hage and Mote, 2008, 2010; Hollingsworth, 2006.

35. These constraints are usually called institutional patterns, and they reflect the major thrust of institutional analysis, which is discussed at the end of Chapter 1. Chapter 4 provides a number of examples of these patterns in the case of French society, while this Introduction points to some of the other factors that have aided Germany and Italy in protecting their industries against globalization.

36. As well as the disciplinary differences and, of course, the cultural ones.

37. Hage, Jordan, and Mote, 2007b; Jordan, Hage, and Mote, 2008.

38. Unlike most social theorists who avoid conducting research, I have always attempted to draw out the theoretical lessons from empirical research and tested theoretical ideas in exacting research designs.

39. Tassey, 2009: 5.

40. Tassey, 2009.

41. Tassey, 2009: 29.

42. And even so, these alarm bells have not been heard by the well-educated public. Most people I know, including many sociologists who study knowledge and innovation, did not know about the deficits in the trade balances of the high-tech sectors. Part of the reason for this ignorance is the continual emphasis on the United States being first in productivity.

43. Dertouzos et al., 1989.

44. Pfeffer, 1981. Dertouzos et al., 1989, provide many examples of this; see pp. 13 and 175 and the strategy of abandoning low-profit, high-volume market segments: 55–56, 257–58.

45. Tassey, 2009: 3–4, who observes that this encourages business elites and policy makers to ignore the major structural changes that are occurring. Dertouzos et al., 1989; 47–49, observe that the mass production system strategy is obsolete.

46. Chandler, 1977.

47. Lafley and Charan, 2008: 296 and 80–81, respectively.

48. Piore and Sabel, 1984.

49. Zammuto and O'Connor, 1992. Evidence for this is provided in Dertouzos et al., 1989: 19–20, 182–84.

50. Duchesneau, Cohn, and Dutton, 1979.

51. Part of the increase in productivity occurs because of learning, but this learning stops; that is, there is upper boundary as to how much additional productivity is possible because of learning how to better employ the machines. Thus only changes in the manufacturing process can maintain increases in productivity.

52. Piore and Sabel, 1984, would also emphasize the importance of network organizations in middle Italy. Their existence rested on a special Italian law that exempted

firms of ten or fewer employees from paying large social security payments. This law encouraged firms to continue to bifurcate when they reached this number. Also see Lazerson, 1993, Dertouzos et al., 1989: 17.

53. Putnam et al., 1993, would also emphasize the importance of social capital, that is, the willingness of individuals and companies to cooperate for the collective good. This theme is important in understanding how to create Silicon Valleys; see Chapter 5.

54. Klepper and Simons, 1997, and Dertouzos et al., 1989: 217–31.

55. Hage et al., 1993. Additional evidence is from companies that change their strategy to combine R&D with flexible manufacturing; see Dertouzos et al., 1989: 14, 17, 237, 301.

56. Jaikumar, 1986.

57. Shaiken, 1984.

58. Personal communication with an engineer in a major machine manufacturer of robots at a conference on flexible engineering, Norfolk, VA, 1984.

59. Jaikumar, 1986.

60. Duchesneau, Cohn, and Dutton, 1979, in their detailed case studies found that engineers were opposed to the new flexible machines whereas the marketing people were in favor. The latter could only win their point of view when the company had decentralized decision making, which is the exception rather than the rule in this industry; see Hage (1999) for a review of the literature on complex social structures in organizations and innovation.

61. Shaiken, 1984.

62. Dankbaar, 1994; Prais, Jarvin, and Wagner, 1989; Steedman and Wagner, 1987, 1989; and for the macro level, see Dertouzos et al., 1989: 17, 20.

63. Streeck et al., 1987.

64. This institutional pattern is called coordinated market economies in distinction to the free market economies associated with the Anglo-Saxon countries such as the United States. These institutional patterns make cooperation easier between the public and the private sectors as well as between labor and management; see Hall and Soskice (2001) and, more generally, the varieties of capitalism.

65. Pfeffer, 1981.

66. Kelley Holland, "Is It Time to Retrain B-Schools?" *New York Times*, Business Section, March 15, 2009.

67. Michael Mandel, "Innovation, Interrupted: How America's Failure to Capitalize on Innovation Hurt the Economy—and What Happens Next, *Business Week*, June 14, 2009.

68. Tassey, 2009: 3 fn 6.

69. This idea is called "spill-overs" and comes from the work of Romer (1994).

70. Jordan, 2006.

71. Nooteboom, 1999; also see Wuyts et al., 2005; Nooteboom et al., 2007.

72. The robots analysis earlier is one example and Tassey (2009) provides many others, which are reported in Chapter 6.

73. Romer (1994) is responsible for this argument and, of course, Silicon Valleys are the prime example. A particularly interesting one in the United States is the marriage between computers and the movie industry.

74. Molas-Gallart and Davies, 2006.

75. Arnold, 2004.

76. Nelson, 1993; see note 28 for a literature that calls into question the idea of the national innovation system and also the evidence reported in Chapter 1.

Chapter 1

1. NSF, 2008: chap. 6.

2. NSF, 2010.

3. Tassey, 2009: 16.

4. Christensen, 1997.

5. Ashlee Vance, "Hewlett-Packard Fights Back with Printers Built for a Digital World, *International Herald Tribune* , June 7, 2010.

6. What economists call externalities. One of the important tasks of research is to determine these.

7. Herrigel, 1994; Steedman and Wagner, 1987, 1989; Dertouzos et al., 1989: 17, 20.

8. Hage and Powers, 1992.

9. Argyris and Schon, 1995, which describes single- and double-loop thinking. What causes single-loop thinking is the cognitive business models discussed extensively in the previous chapter. Chapter 5 provides examples of some companies, IBM and Ericsson, that broke out of their typical business model, especially regarding intellectual property.

10. Hage and Powers, 1992.

11. Pine, 1993.

12. Penn and Zalesne, 2007.

13. Mote, Jordan, and Hage, 2007.

14. Connie Guglielmo, "Apple Raises Heat on Android Rivals," *International Herald Tribune*, July 9, 2010.

15. Mote, Jordan, and Hage, 2007.

16. Powell and Brantley, 1992; Powell, Koput, and Smith-Doerr, 1996; Powell, 1998.

17. Pernish, 2009: 162–63.

18. Oosterwijk in van Waarden et al., 2003, pt. 1: 20–23; in particular read the examples in the drug industry.

19. The first time these ideas were developed was for a conference on science and technology policy; see Hage, Jordan, and Mote, 2007b.

20. Casper, 2006.

21. This is called tacit knowledge; Polanyi, 1966.

22. Chesbrough, 2006.

23. This thesis is called Mode 2; Gibbons et al., 1994.

24. Blau, 1970; Hage, 1980.

25. Chesbrough, 2006.

26. Gomes-Casseres, 1996.

27. Kline and Rosenberg, 1986.

28. Chesbrough, 2006. The same is true of the idea of absorptive capacity, which is simply measured by the amount of RDT in the organization; see Cohen and Levinthal, 1990.

29. Sawyer, 2007: 159.

30. Lafley and Charan, 2008: 143–44.

31. The U.S. culture is one that encourages risk taking but at the same time discourages cooperation either between private firms or between the public and the private sector. In contrast, the European cultures do not have risk-taking cultures, but they do encourage cooperation. This flows from the Anglo-Saxon liberal market thinking and from the associational market reasoning; see Hall and Soskice, 2001.

32. Van Waarden and Oosterwijk, 2006.

33. For the project reports see van Waarden et al., 2003.

34. Oosterwijk in van Waarden et al., 2003, pt. 1: 14–19.

35. Although the pattern was the opposite in the United States where a private company was broken into the seven little "bells," in Europe, public monopolies began to have competition.

36. Van Waarden et al., 2003.

37. Anderson and Tushman, 1990.

38. Van Waarden and Oosterwijk, 2006.

39. This has two important implications: The national system of innovation model is less relevant, and it is critical to examine each sector separately rather than together.

40. Powell, 1998.

41. Casper, 2006.

42. A major exception is Tassey, 2009.

43. A recent article in *Science* states that the predictions about the impact of global warming have gone from bad to worse; see Kintisch, 2009: 1546–47. Of course, not everyone agrees with the various models that are used to make these predictions. It is true that there are cycles of warming and cooling, and some believe that we are merely in another cycle. At the same time, the level of carbon dioxide in the ocean implies that this cycle may be much worse because of man. Even if one does not accept the models, and there are many problems and complexities that they do not handle well including the short-term carbon cycles, one can agree that it is desirable to reduce

carbon footprints for the benefit of Earth and certainly reduce the dependency on fossil fuels and oil in particular.

44. John Broder, "U.S. Needs to Catch Up on Energy Research, Executives Warn," *International Herald Tribune,* June 11, 2010.

45. Seria, 2009: 1257–59.

46. Ohlrogge et al., 2009: 1019–20.

47. Charles, 2009: 172–75.

48. Finkel, 2009: 1004–5.

49. Normile, 2009: 1260–61.

50. Corvin, 2007.

51. Examples of this line of reasoning are Hollingsworth, 1997, and Whitely, 1992; the former called his model technical systems, while the latter labeled essentially the same idea as business systems.

52. Nelson, 1993.

53. Hollingsworth, 1997; Hall and Soskice, 2001.

54. Hall and Soskice, 2001.

55. Hollingsworth, 2006; Levi, 1997; Pierson, 2000: 252.

56. Argyris and Schon, 1995.

57. Campbell, 2004.

Chapter 2

1. Brown and Eisenhardt, 1995; Hage, 1999; Kanter, 1988. Technically, most of my earlier work was concerned with the complex organization rather than the complex research team. Only when I began to work on teams in science and technology in the 1990s did I apply the same set of ideas to research teams, as is indicated in Chapter 4. Also see Hollingsworth, 2006, a framework that we developed together.

2. Sawyer, 2007: 18–19.

3. Cummings and Kiesler, 2005: 710.

4. Stokols et al., 2008; other important references in this literature are Klein, 2008, and Wuchty, Jones, and Uzzi, 2007.

5. Shenhar, 1999, 2001.

6. Gina Kolata, "Playing It Safe in Cancer Research," *New York Times*, June 28, 2009.

7. Manton et al., 2009, reports this progress. Over time, it became clear that some genes do not cause cancer but regulate the growth process of cancer. To make matters more complicated, each cancer appears to have different subtypes associated with distinctive genes, to say nothing about the variety of external causes, which vary from one cancer to the next. The context of the cell has become an exciting new avenue of research.

8. Jordan, 2006; also see Jordan et al., 2005; Jordan, Streit, and Binkley, 2003a and b.

9. These ideas are the most exciting aspects of Jordan's (2006) theory of research profiles.

10. The reader may be surprised by the amount of detail in this definition, but in a large-scale research project funded by the NSF, Gretchen Jordan, Jon Mote, and I have discovered that there is a considerable variety in the ways in which projects are designed in the national laboratories and even in the terms used to describe them. Also, the role of networks, that is, consulting and conferring with other experts, sometimes makes drawing the boundaries of specific projects somewhat difficult. Indeed, this is one of the reasons why network coordination for radical innovations is so important. My colleague Jonathan Mote is presently writing a paper on this important and previously unrecognized problem.

11. Stokols et al., 2008: S79.

12. Corvin, 2007.

13. Sawyer, 2007: 109.

14. Cummings and Kiesler, 2007: 1623.

15. Cummings and Kiesler, 2005: 703.

16. I appreciate that immediately the reader will start to think: What about Einstein? As Sawyer (2007) observes, Einstein relied upon a large accumulation of physical data collected in various laboratories to develop many of his seminal ideas. Most scientific breakthroughs in biomedicine, except for a few in the nineteenth century and early part of the twentieth century, were developed by teams (Hage and Mote, 2010; Hollingsworth, 2006; also see the large list of examples in Stokols et al., 2008). Furthermore, it is much easier for an individual to develop new theories than to conduct research that involves empirical findings.

17. Lafley and Charan, 2008: 296.

18. Stokols et al., 2008.

19. Mote, Jordan, and Hage, 2007.

20. Kohler, 1994.

21. Allen and Cohen, 1969.

22. Hage and Mote, 2008; see also Chapter 4.

23. Dougherty, 1992. Also see Dougherty and Corse, 1996 (cited in Brown and Eisenhardt, 1995: 357).

24. Lawrence and Lorsch, 1967.

25. Mendelsohn, 1996: 115–16.

26. This idea of measuring the diversity of the research team by comparing which journals and other sources team members read originated in a very interesting article by Rafols and Meyer, 2010. Their concern was measuring the diversity in diverse interdisciplinary teams such as those in nanotechnology.

27. Judson, 1979.

28. Nor does the winning of the Nobel Prize by Watson and Crick deny the importance of a measuring instrument, crystallography, in the development of the insights for the description of the double helix, another measure of diversity and of the importance of research technology.

29. Judson, 1979.

30. Hollingsworth, 2006; Hage and Mote, 2010.

31. In the beginning of the century that is involved in this study, some of the discoveries where made by single individuals, but across time, this become more and more rare. Furthermore, the team consisted of more individuals than those recognized by the prize. Hollingsworth and Hage used the prize simply as a means of identifying where the work was accomplished so that they could isolate transformational organizations for research purposes (see Chapter 4).

32. Cressac, 1950; Lagrange, 1954.

33. Mote, Jordan, and Hage, 2007.

34. Mote, Jordan, and Hage, 2007.

35. Stokols et al., 2008.

36. In economics, when some team members do not contribute as much effort as others to team performance, it is called the free rider problem. In Chapter 3, the importance of having a collective vision that motivates people is discussed. This idea of a collective vision partially speaks to this problem.

37. It would take us too far afield to indicate all the structural properties that can impact on the reward system as other factors that can affect the performance of individuals within the research. Centralized organizations, with a rigid status system, and many bureaucratic rules are likely to present a number of difficulties for managers interested in developing a diverse research team; see Hage, 1980. For this reason, Chapter 7 focuses on blockages in organizations.

38. Pelz and Andrews, 1976.

39. Judge, Fryxell, and Dooley, 1997: 77.

40. Jordan, 2006.

41. Gelès et al., 2000: 18.

42. The selection of these cutoffs may appear to be somewhat arbitrary. And $10 million seems high to researchers who obtain small grants from NSF that do not pay for personnel. The $10 million assumes the cost of personnel, the overhead, and a project time of several years, if not more.

43. Gelès et al., 2000.

44. Shenhar, 2001.

45. Judge, Fryxell, and Dooley, 1997: 80.

46. Jordan, 2005; Hage, Jordan, and Mote, 2008.

47. Hage, Jordan, and Mote, 2008.

48. Sicotte and Langley, 2000.

49. Brown and Eisenhardt, 1995: 358.

50. Brown and Eisenhardt, 1995.

51. Chesbrough, 2006.

52. Lafley and Charan, 2008: 80–81.

53. Mote, Jordan, and Hage, 2007, argue in a review of the network literature that this is major problem.

54. Among other references, the reader might find the following useful: Alter and Hage, 1993; Dittrich, Duysters, and de Maro, 2007; Dussauge and Garrette, 1999; Kogut, Shan, and Walker, 1993; Lundvall, 1992; O'Doherty, 1995; Powell, 1998. Most of this literature has focused on a single kind of research organization such as a university or a firm and a single kind of linkage, either joint ventures or alliances of one kind or another. They have not focused on the six arenas within the technical sector.

55. Chesbrough, 2006; chapter 9 in Chesbrough discusses a number of strategies that in various ways imply some of the arenas, but they are not explicit. In particular, there is no discussion of manufacturing and quality research as critical arenas.

56. Cohen and Levinthal, 1990, introduced the concept of absorptive capacity measured by the amount of money spent on RDT. The diversity of funded arenas is more important, as is the presence of technological gatekeepers.

57. The success of the U.S. Department of Agriculture in helping farmers achieve incredible yields via the extension services located at a number of the state universities during the past century has been largely forgotten, but they were able to increase our agricultural yields by 4 percent per annum. The same growth rates in industry would contribute considerably to restoring our trade balances; see Ferleger and Lazonick, 1993.

58. However, the growing complexity of problems has increasingly forced universities to seek expertise elsewhere, as is indicated in Chapter 5.

59. Specific examples are in the sociology of science (Jordan, 2006; Hollingsworth, 2006), in the R&D management literature (Brown and Eisenhardt, 1995; Hage, 1999), in the new team science literature (Stokols et al., 2008; Rafols and Meyer, 2010), and in the project management literature (Shenhar, 2001).

60. Similar ideas are to be found in Shenhar, 2001.

Chapter 3

1. These kinds of information define the concept of tacit knowledge (Polanyi, 1966). His original model was the athlete who learns through practice how to improve his performance but cannot explain why a particular position or posture works better. But this same idea can and has been applied to the microcosm of the research team. Not unexpectedly, the very best work on this is Japanese; see Nonaka and Takeuchi, 1995.

2. Definitions of these concepts can be found in Hage and Meeus (2006: Chap. 1).

3. Also see Nooteboom et al., 2007; Wuyts et al., 2005.

4. And Nooteboom has evidence to support the idea that this is in fact what organizations do; see Wuyts et al., 2005; Nooteboom et al., 2007. These references also connect these ideas with absorptive capacity, perceiving "optimal cognitive distance" as a narrow range.

5. Rafols and Meyer, 2010.

6. Hage and Mote, 2008 and 2010; Hollingsworth, 2006.

7. The various references to the Pasteur Institute represent work with Jonathan Mote (see Hage and Mote, 2008 and 2010). The extended case study that forms the third part of this chapter is research work in a national laboratory; see Hage, Jordan, and Mote, 2007a.

8. Gascar, 1986: 135; Faure, 1988.

9. Mendelsohn, 1996: 111.

10. Delaunay, 1962: 161.

11. Lafley and Charan, 2008.

12. Judge, Fryxell, and Dooley, 1997: 75.

13. Legout, 1999.

14. Lawrence and Lorsch, 1967.

15. Much of this debate focuses on the idea that basic science will be corrupted if it has an applied focus. Yet, all the evidence is to the contrary. Admittedly, it is sometimes difficult to see the applied side of high-energy physics, but this does not mean that it would be wasted time to look for applied advantages to even the most basic of scientific research programs.

16. J. R. Hollingsworth, personal communication, 1998–99.

17. Hage and Mote, 2008.

18. The thrust of Nooteboom's work (1999: 14; Wuyts et al., 2005; Nooteboom et al., 2007) is to develop a cognitive model for organizations.

19. Polanyi, 1966; Nonaka and Takeuchi, 1995; Levin and Cross, 2004.

20. Mohrman, Galbraith, and Monge, 2004; Hage, Jordan, and Mote, 2007a.

21. Bureaucracy is another one of those areas in which there are multiple explanations for the existence of rules and various other blockages that are related to it; see Hage, 1980. Perhaps the strongest evidence for a negative impact on innovation is the routinization of work.

22. Judge, Fryxell, and Dooley, 1997.

23. The learning was measured in this study by the recentness of the citations in their patents.

24. Hage, 1974.

25. Seidell, 1926.

26. Kidder, 1981.

27. Judge, Fryxell, and Dooley, 1997.

28. Quoted in Moulin, 1994; the source of reference is reported in Bernard, 1955: n75, in Moulin, 1994.

29. Elie Wolman, personal interview, Pasteur Institute, Paris, 1998.

30. Hage and Powers, 1992.

31. Morgan Scott Peck, in Judge, Fryxell, and Dooley, 1997: 75.

32. Providing for professional development was the practice, but it may not be now. Also, IBM used to guarantee employment, which also developed loyalty. And this policy has been abandoned.

33. George Cohen, personal interview, Pasteur Institute, Paris, 1998.

34. Judson, 1979.

35. Bernard and Nègre, 1939: 50–56.

36. Bernard and Nègre, 1939: 50–56.

37. Hage, Jordan, and Mote, 2007a, 2008.

38. This extended example is taken from Sawyer, 2007: 4.

39. The concept of human capital also needs the concept of human capital decay.

40. Christensen, 1997.

41. Brown and Duguid, 1998; Cohen and Sproull, 1995; Conner and Prahalad, 1996; Grant, 1996.

42. Grant (1996) is one of the few to recognize this problem.

43. Kim and Wilemon, 2007.

44. Actually, the process of evolution and its impact on sociology is much more profound than the creation of new sections. Within sociology, as organizational structures have gradually disappeared with the emphasis on flat organizations and individual autonomy, the topics of culture, emotions, and networks have become important ones, precisely because they represent ways of integrating individuals in the absence of structure (see Hage and Powers, 1992). The theme of culture is represented in the discussion of the importance of a vision, a theme that reemerges in Chapter 4. But perhaps what is more critical is the importance of emotions and practices that encourage bonding between team members.

45. Hage and Powers, 1992.

Chapter 4

1. "Energiser Money," *The Economist,* March 28, 2009.

2. Hage and Mote, 2010.

3. This section is a substantial rewrite of Hage and Mote, 2010.

4. Mendelsohn, 1996: 164.

5. Stokes, 1997.

6. Faure, 1988; Moulin, 1994.

7. The doctrine that the functions of the body are due to a principle different from biochemical reactions and that life is in some respects self-determining.

8. Liebenau and Robson, 1991; Moulin and Guenel, 1993.

9. Pelis, 1995.

10. Cressac, 1950: 118; Debru, 1991; Robinson, 1906.

11. Lépine, 1966: 59, 100.

12. Cressac, 1950.

13. Delaunay, 1962: 111.

14. Delaunay, 1962: 100, 157.

15. Delaunay, 1962: 130, 158.

16. Delaunay, 1962.

17. Moulin, 1994.

18. Lagrange, 1954.

19. Debru, 1991: 107–17; Moulin and Guenel, 1993; Weindling, 1991: 143–45.

20. Cressac, 1950: 133.

21. Gascar, 1986: 179–80.

22. Mery, 1987.

23. Hage, 1980; Mohrman, Galbraith, and Monge, 2004.

24. This section is a substantial rewrite of Hage and Mote, 2008.

25. Bourdieu and Passeron, 1977.

26. Pasteur Institute, 1911.

27. Dedet, 2000; Léonard, 1981: 250.

28. Huguet, 1991.

29. Charle and Telkes, 1989.

30. Charle and Telkes, 1989; Weisz, 1995.

31. Charle and Telkes, 1988.

32. Crosland, 1992.

33. Delaunay, 1962.

34. Bass and Riggio, 2006.

Chapter 5

1. Jason Dedrick, Kenneth Kraemer, and Greg Linden, "Who Profits from Innovation in Global Value Chains? A Study of the iPod and Notebook PCs," paper presented at the annual conference of the Alfred P. Sloan Foundation, May 1–2, 2008, Boston.

2. Dedrick, Kraemer, and Linden, "Who Profits."

3. Block and Keller, 2009.

4. Cummings and Kiesler, 2005.

5. Actually, proprietary issues are now more common in public universities as they seek to find additional sources of funding to replace declining state support.

6. Cummings and Kiesler, 2005: 707–9.

7. Cummings and Kiesler, 2005: 718.

8. Cummings and Kiesler, 2005: 710–11.

9. Chesbrough, 2006; Dittrich, Duysters, and de Maro, 2007.

10. Utterback, 1994.

11. Chesbrough, 2006: 99–101.

12. Chesbrough, 2006: 99–101.

13. Chesbrough, 2006: 101–4.

14. Dittrich, Duysters, and de Maro, 2007: 1496–1511.

15. Branscomb, 2008.

16. Tassey, 2009: 32.

17. Womack, Jones, and Roos, 1990. The material for this discussion comes from the book *The Machine That Changed the World,* which is based on the $5 million MIT study of the automobile industry worldwide. Although the study is twenty years old, many of the basic principles are still relevant.

18. Clark, Fujimoto, and Chew, 1987, in Womack, Jones, and Roos, 1990: 118.

19. Womack, Jones and Roos, 1990: chap. 5.

20. Clark, Fujimoto, and Chew, 1987, in Womack, Jones, and Roos, 1990: 118.

21. Nonaka and Peltokorpi, 2006.

22. Womack, Jones, and Roos, 1990: 11, 60–68.

23. Brown and Eisenhardt, 1995: 361–62.

24. Sánchez and Pérez, 2003. In my opinion, they made a mistake by doing factor analysis rather than focusing on the specific managerial practices in combination with cooperation that were most highly associated with quick product development times and lower costs.

25. Browning, Beyer, and Shelter, 1995: 114.

26. Other possible explanations are the depression in Japan, the rise of the Korean chip maker, Samsung, and U.S. tariff restrictions. But these alternative explanations are not convincing because this is an international product and the new manufacturing techniques developed by SEMATECH clearly made the American manufacturers more competitive.

27. Carayannis and Alexander, 2004.

28. Browning, Beyer, and Shelter, 1995: 119.

29. Carayannis and Alexander, 2004: 226.

30. Browning, Beyer, and Shelter, 1995: 117.

31. Browning, Beyer, and Shelter, 1995: 124.

32. Browning, Beyer, and Shelter, 1995: 117.

33. Browning, Beyer, and Shelter, 1995: 123.

34. Browning, Beyer, and Shelter, 1995: 126.

35. Carayannis and Alexander, 2004: 231.

36. Carayannis and Alexander, 2004: 229.

37. Carayannis and Alexander, 2004: 229.

38. Carayannis and Alexander, 2004: 229.

39. Tassey, 2009: 28.

40. Tassey, 2009: 32.

41. Casper, 2006.

42. Matthews, 1997.

43. Tassey, 2009: 30.

44. Kao, 2007.

45. Kao, 2007. The fact that this company could transform itself from a wood pulp producer into the technological leader in cell phones is an objective lesson for all American CEOs. The only comparable case in the United States is the movement of Studebaker out of covered wagons into automobiles.

46. Hall and Soskice, 2001.

Chapter 6

1. Tassey, 2009: 30.

2. Tassey, 2009: 27.

3. Tassey, 2009: 35–37.

4. Tassey, 2009: 30.

5. Jeffrey R. Immelt, interview with Charlie Rose, Public Broadcasting System, June 25, 2009.

6. Steven Greenhouse, "GE to Add Two New U.S. Plants as Unions Agree on Cost Controls," *New York Times,* August 7, 2009. As this book goes to press, GE has announced new initiatives to expand manufacturing in the United States, including energy-efficent washers and dryers, sodium batteries, and so on. See Steve Lohr, "GE Goes with What It Knows: Making Stuff," *New York Times*, December 5, 2010.

7. Much of this first section has been inspired by Tassey, 2009.

8. Tassey, 2009: 9.

9. Tassey, 2009: 18–20.

10. Tassey, 2009: 6.

11. This is called the *economies of scale argument.*

12. Tassey, 2009: 24–25.

13. Tassey, 2009: 25.

14. Block and Keller, 2009. The averages were computed based on their data in appendix 1 by adding together the number of awards given across three years or six and then divided by the number of awards made within these years, subtracting out those awarded overseas.

15. National Renewable Energy Laboratory, *Energy Innovations* (Winter 2010): 5.

16. "Energiser Money," *The Economist*, March 28, 2009.

17. Brookhaven National Laboratory, *Research for Our Energy Future,* n.d.: 7.

18. Block and Keller, 2009: 472–73.

19. Block and Keller, 2009: 473. It could have been sooner but because this research study measures only three years in each decade, it might have missed the first time.

20. Block and Keller, 2009: 477.

21. U.S. Department of Commerce, *Measuring ATP Impact: 2006 Report on Economic Progress,* Washington, DC: U.S. Department of Commerce.

22. For the following cases and others, see U.S. Department of Commerce, *Measuring ATP Impact,* 27–33.

23. Whitley, 1992.

24. Hall and Soskice, 2001; Hollingsworth, 1997; Hollingsworth and Boyer, 1997; Whitley, 1994.

25. Christensen, 1997.

26. Tassey, 2009.

27. Ferleger and Lazonick, 1993, which argues how state then helped develop the country.

28. Rassmussen, 1989: 120.

29. Putnam, 1993; his concept of social capital was developed first with an intensive study of the middle regions of Italy. The Midwest of the United States has had a considerable amount of social capital, and I would argue one of the reasons for this is because of the actions of the USDA.

30. Greenhouse, "G.E. to Add Two New U.S. Plants."

31. NSF's science and engineering indicators can be found at http://www.nsf.gov/statistics.

32. Hage and Hollingsworth, 2000; Nieminen and Kaukonen, 2001.

33. Another advantage of collecting data on the technical progress in each of the arenas is that it provides the missing link between short-term evaluations discussed in Chapter 7 and the medium- and long-term ones typical at the societal level. This also speaks to two problems in systemic evaluations: (1) the tendency is for system evaluations to be conducted over the medium and the long terms, but the requirement is for quick feedback with quantifiable indicators for the benefit of policy makers; and (2) the need to identify the gaps or weak connections in the idea innovation network.

34. Beyond this, by establishing the missing link between short- and medium- or long-term evaluations at the systemic level, one is also constructing a theory of science of science and innovation policy, a current objective of NSF, and developing a number of insights about blockages and obstacles located in institutional arrangements. Thus this extensive data collection can make the social sciences stronger just as the federal government collection of productivity data has helped strengthen economics.

Chapter 7

1. U.S. Department of Commerce, *Measuring ATP Impact: 2006 Report on Economic Progress* (Washington, DC: U.S. Department of Commerce); see appendix.

2. Morton et al., 2006.

3. Lu et al., 2009.

4. Dudley et al., 2005. This treatment by customized vaccines is still in the experimental stage and the reports of numbers are based on very few patients and really more hypothetical at this point.

5. Grau et al., 2009.

6. Andersen et al., 2008.

7. This feedback is part of a larger framework; see Hage, Jordan, and Mote, 2007b; Jordan, Hage, and Mote, 2008.

8. Jordan, 2006; also see Jordan and Streit, 2003; Jordan, Streit, and Binkley, 2003a, 2003b.

9. Woodman, Sawyer, and Griffin, 1993.

10. Brown and Eisenhardt, 1995: 364.

11. Brown and Eisenhardt, 1995.

12. Judge, Fryxell, and Dooley, 1997: 80.

13. J. Mote and J. Hage, "Research as Usual? A Longitudinal Study of Organizational Change in R&D" (paper, Center for Innovation, University of Maryland, 2010).

BIBLIOGRAPHY

Afuah, A. N., and Bahram, N. (1995). "The Hypercube of Innovation." *Research Policy* **24**: 51–76.

Allen, T. J., and Cohen, S. I. (1969). "Information Flow in Research and Development Laboratories." *Administrative Science Quarterly* **14** (1): 12–19.

Alter, C., and Hage, J. (1993). *Organizations Working Together.* Newbury Park, CA: Sage.

Andersen, B., et al. (2008). "Psychologic Intervention Improves Survival for Breast Cancer Patients: A Randomized Clinical Trial." *Cancer* **113** (12): 3450–58.

Anderson, P., and Tushman, M. L. (1990). "Technological Discontinuities and Dominant Designs: A Cyclical Model of Technological Change." *Administrative Science Quarterly* **35** (4): 604–33.

Argyris, C., and Schon, D. (1995). *Organizational Learning II: Theory, Method and Practice.* Upper Saddle River, NJ: Prentice-Hall.

Arnold, E. (2004). "Evaluating Research and Innovation Policy: A Systems World Needs Systems Evaluations." *Research Evaluation* **13** (1): 3–17.

Augustine, N. R. (2005). *Rising Above the Gathering Storm: Energizing and Employing America for a Brighter Economic Future.* Washington, DC: National Academy of Science Press.

Bass, B., and Riggio, R. (2006). *Transformational Leadership*, 2d ed. Mahwah, NJ: Erlbaum.

Bernard, P. N., and Négre, L. (1939). *Albert Calmette: Sa vie, son oeuvre scientifique.* Paris: Masson.

Blau, P. M. (1970). "A Formal Theory of Differentiation in Organizations." *American Sociological Review* **35** (2): 201–18.

Block, F., and Keller, M. (2009). "Where Do Innovations Come From? Transformations in the U.S. National Innovation System, 1970–2006." *Socio-Economic Review* **7** (3): 459–83.

Bourdieu, P., and Passeron, J. C. (1977). *Reproduction in Education, Society and Culture.* London: Sage.

Branscomb, L. (2008). "Research Alone Is Not Enough." *Science* **323**: 915–16.

Brown, J., and Duguid, P. (1998). "Organizing Knowledge." *California Management Review* **40** (3): 90–113.

Brown, S. L., and Eisenhardt, K. M. (1995). "Product Development: Past Research, Present Findings and Future Directions." *Academy of Management Review* **20** (2): 343–78.

Browning, L. D., Beyer, J., and Shelter, J. (1995). "Building Cooperation in a Competitive Industry: Sematech and the Semiconductor Industry." *Academy of Management Journal* **38** (1): 113–51.

Campbell, J. L. (2004). *Institutional Change and Globalization*. Princeton, NJ: Princeton University Press.

Carayannis, E. G., and Alexander, J. (2004). "Strategy, Structure, and Performance Issues of Precompetitive R&D Consortia: Insights and Lessons Learned from Sematech." *IEEE Transactions on Engineering Management* **51** (2): 226–32.

Casper, S. (2006). "Exporting the Silicon Valley to Europe: How Useful Is Comparative Institutional Theory?" In J. Hage and M.T.H. Meeus, eds., *Innovation, Science and Institutional Change: A Research Handbook*, 483–504. London: Oxford Press.

Chandler, J. A. (1977). *The Visible Hand: The Managerial Revolution in American Business*. Cambridge, MA: Belknap Press.

Charle, C., and Telkes, E. (1988). *Les professeurs du Collège de France: Dictionnaire biographique, 1901–1939*. Paris: National Institute for Educational Research.

Charle, C., and Telkes, E. (1989). *Les professeurs de la Faculté des Sciences de Paris, Dictionnaire biographique, 1901–1939*. Paris: National Institute for Educational Research.

Charles, D. (2009). "Renewables Test I.Q. of the Grid. " *Science* **324**: 172–75.

Chesbrough, H. (2006). *Open Innovation*. Boston: Harvard Business School.

Christensen, C. (1997). *The Innovator's Dilemma*. Boston: Harvard Business School.

Cohen, M. D., and Sproull, L. S. (1995). *Organizational Learning*. Newbury Park, CA: Sage.

Cohen, W. M., and Levinthal, D. A. (1990). "Absorptive Capacity: A New Perspective on Learning and Innovation." *Administrative Science Quarterly* **35** (1): 128–52.

Committee Members of the 2005 Rising Above the Gathering Storm Report (2010). *Rising Above the Gathering Storm, Revisited: Rapidly Approaching Category 5*. Washington, DC: The National Academies Press.

Conner, K. R., and Prahalad, C. K. (1996). "A Resource-Based Theory of the Firm: Knowledge Versus Opportunism." *Organization Science* **7** (5): 477–501.

Corvin, A. (2007). "Metal of the Future?" *Washington CEO* (November): 46–49.

Cressac, M. (1950). *Le Docteur Roux: Mon oncle*. Paris: L'Arche.

Crosland, M. P. (1992). *Science Under Control: The French Academy of Sciences, 1795–1914*. Cambridge: Cambridge University Press.

Cummings, J., and Kiesler, S. (2005). "Coordination Costs and Project Outcomes in Multiple University Collaborations." *Social Studies in Science* **35** (5): 703–22.

Cummings, J., and Kiesler, S. (2007). "Coordination Costs and Project Outcomes in Multiple University Collaborations." *Research Policy* **36**: 1620–34.

Dankbaar, B. (1994). "Sectorial Governance in the Automobile Industries of Germany, Great Britain, and France." In J. R. Hollingsworth, P. C. Schmitter, and W. Streeck, eds., *Governing Capitalist Economies: Performance and Control of Economic Sectors*, 156–82. Oxford: Oxford University Press.

Debru, C. (1991). "Actualité d'Emile Duclaux." In M. Morange, *L'Institut Pasteur: Contributions à son histoire*, 107–14. Paris: La Découverte.

Dedet, J.-P. (2000). *Les Instituts Pasteur d'Outre-Mer*. Paris: L'Harmattan.

Delaunay, A. (1962). *L'Institut Pasteur des Origines à Nos Jours*. Paris: France-Empire.

Dertouzos, M., et al. (1989). *Made in America: Regaining the Productive Edge*. Cambridge, MA: The MIT Press.

Dittrich, K., Duysters, G., and de Maro, A. D. (2007). "Strategic Repositioning by Means of Alliance Networks: The Case of IBM." *Research Policy* **36** (10): 1496–1511.

Dougherty, D. (1992). "Interpretive Barriers to Successful Product Innovation in Large Firms." *Organization Science* **3** (2): 179–202.

Dougherty, D., and Corse, S. (1996). "What Does It Take to Take Advantage of Product Innovation?" In D. Rosenbloom, and R. Burgelman, eds., *Research on Technological Innovation, Management and Policy*, 6: 424–39. Greenwich, CT: JAI Press.

Duchesneau, T., Cohn, S., and Dutton, J. (1979). Case Studies of Innovation Decision-Making in the U.S. Footwear Industry. Manuscript. Orono: University of Maine.

Dudley, M., et al. (2005). "Adoptive Cell Transfer Therapy Following Non-Myeloablative but Lymphodepleting Chemotherapy for the Treatment of Patients with Refractory Metastatic Melanoma." *Journal of Clinical Oncology* **23** (10): 2346–57.

Dussauge, P., and Garrette, B. (1999). *Cooperative Strategy: Competing Successfully Through Strategic Alliances*. Chichester, NY: Wiley.

Estrin, J. (2008). *Closing the Innovation Gap: Reigniting the Spark in a Global Economy*. New York: McGraw-Hill.

Faure, M. (1988). *Histoire des cours de l'Institut Pasteur*. Paris: Pasteur Institute Archives.

Ferleger, L., and Lazonick, W. (1993). "The Managerial Revolution and the Developmental State: The Case of U.S. Agriculture." *Business and Economic History* **22** (3) (Winter).

Finkel, E. (2009). "Making Every Drop Count in the Buildup to a Blue Revolution." *Science* **323**: 1004–5.

Gascar, P. (1986). *Du côté de chez Monsieur Pasteur*. Paris: Odile Jacob.

Gelès, C., et al. (2000). *Managing Science: Management for R and D Laboratories*. New York: Wiley.

Gibbons, M., et al. (1994). *The New Production of Knowledge: The Dynamics of Science and Research in Contemporary Societies*. London: Sage.

Gomes-Casseres, B. (1996). *The Alliance Revolution: The New Shape of Business Rivalry*. Cambridge, MA: Harvard University Press.

Grant, R. M. (1996). "Toward a Knowledge-Based Theory of the Firm." *Strategic Management Journal* **17**: 109–22.

Grau, M. V., et al. (2009). "Nonsteroidal Anti-Inflammatory Drug Use After 3 Years of Aspirin Use and Colorectal Adenoma Risk: Observational Follow-up of a Randomized Study." *Journal of the National Cancer Institute* **101**: 267–76.

Guerrieri, P., and Tylecote, A. (1998). "Cultural and Institutional Determinants of National Technical Advantage." In R. Coombs, K. Green, and V. Walsh, eds., *Technological Change and Organization*, 180–209. Aldershot, UK: Elgar.

Hage, J. (1974). *Communication and Organizational Control: Cybernetics in Health and Welfare Settings*. New York: Wiley.

Hage, J. (1980). *Theories of Organizations: Form, Process, and Transformation*. New York: Wiley.

Hage, J. (1999). "Organizational Innovation and Organizational Change." *Annual Review of Sociology* **25** (1): 597–622.

Hage, J., and Hollingsworth, J. R. (2000). "A Strategy for the Analysis of Idea Innovation Networks and Institutions." *Organization Studies* **21** (5): 971–1004.

Hage, J., Jordan, G. B., and Mote, J. (2007a). *R&D Integration: How to Build a Diverse and Integrated Knowledge Community*. 2007 Annual Conference of Portland International Center for Management of Engineering and Technology (PICMET), Portland, OR.

Hage, J., Jordan, G., and Mote, J. (2007b). "A Theories-Based Systematic Framework for Evaluating Diverse Portfolios of Scientific Work, Pt. 2: Macro Indicators and Policy Interventions." *Science and Public Policy* **34** (10): 731–41.

Hage, J., Jordan, G. B., and Mote, J. (2008). "Designing and Facilitating R&D Collaboration: The Balance of Diversity and Integration." *Journal of Engineering and Technology Management* **25** (4): 256–64.

Hage, J., and Meeus, M., eds. (2006). *Innovation, Science and Institutional Change: A Research Handbook*. Oxford: Oxford University Press.

Hage, J., and Mote, J. (2008). "Transformational Organizations and Institutional Change: The Case of the Institut Pasteur and French Science." *Socio-Economic Review* **6** (2): 313–36.

Hage, J., and Mote, J. (2010). "Transformational Organizations and a Burst of Scientific Breakthroughs: The Institut Pasteur and Biomedicine, 1889–1910." *Social Science History* **34** (1): 13–46.

Hage, J., and Powers, C. (1992). *Post-Industrial Lives: Roles and Relationships in the 21st Century*. Newbury Park, CA: Sage.

Hage, J., et al. (1993). "The Impact of Knowledge on the Survival of American Manufacturing Plants." *Social Forces* **72**: 223–46.

Hage, J., et al. (2008). Research Work and Work Satisfaction: The Case of the STAR Research Division in NOAA. Paper. Center for Innovation, University of Maryland, College Park, MD.

Hall, P. A., and Soskice, D., eds. (2001). *Varieties of Capitalism: The Institutional Foundations of Comparative Advantage*. New York: Oxford University Press.

Henderson, R. M., and Clark, K. (1990). "Architectural Innovation: The Reconfiguration of Existing Product Technologies and the Failure of Established Firms." *Administrative Science Quarterly* **35** (1).

Herrigel, G. (1994). "Industry as a Form of Order: A Comparison of the Historical Development of the Machine Tool Industries in the United States and Germany." In J. R. Hollingsworth, P. Schmitter, and W. Streeck, eds., *Governing Capitalist Economies: Performance and Control of Economic Sectors*, 97–128. Oxford: Oxford University Press.

Hollingsworth, J. R. (1997). "Continuities and Changes in Social Systems of Productions: The Cases of Japan, Germany, and the United States." In J. R. Hollingsworth and R. Boyer, eds., *Contemporary Capitalism: The Embeddedness of Institutions*, 265–310. New York: Cambridge University Press.

Hollingsworth, J. R. (2006). "A Path Dependent Perspective on Institutional and Organizational Factors Shaping Major Scientific Discoveries." In J. Hage and M.T.H. Meeus, eds., *Innovation, Science, and Institutional Change: A Research Handbook*, 423–42. London: Oxford University Press.

Hollingsworth, J. R., and Boyer, R., eds. (1997). *Contemporary Capitalism: The Embeddedness of Institutions*. New York: Cambridge University Press.

Huguet, F. (1991). *Les professeurs de la Faculté de Médecine de Paris: Dictionnaire biographique, 1794–1939*. Paris: National Institute for Educational Research.

Jaikumar, R. (1986). "Postindustrial Manufacturing." *Harvard Business Review* **64**: 69–76.

Jordan, G. B. (2005). "What Is Important to S&T Workers." *Research Technology Management* **48** (3): 23–32.

Jordan, G. B. (2006). "Factors Influencing Advances in Basic and Applied Research: Variation Due to Diversity in Research Profiles." In J. Hage and M.T.H. Meeus, eds., *Innovation, Science, and Institutional Change: A Research Handbook*, 173–95. London: Oxford University Press.

Jordan, G. B., Hage, J. T., and Mote, J. (2008). "A Theories-Based Systemic Framework for Evaluating Diverse Portfolios of Scientific Work, Pt. 1: Micro and Meso Indicators." *Reforming the Evaluation of Research: New Directions for Evaluation* **118**: 7–24.

Jordan, G. B., and Streit, L. D. (2003). "Recognizing the Competing Values in Science and Technology Organizations: Implications for Evaluation." In P. Shapira and S. Kuhlmann, eds., *Learning from Science and Technology Policy Evaluations*. Northampton, MA: Elgar.

Jordan, G. B., Streit, L. D., and Binkley, J. (2003a). "Assessing and Improving the Effectiveness of National Research Laboratories." *IEEE Transactions on Engineering Management* **50** (2): 228–35.

Jordan, G., Streit, L. D., and Binkley, J. (2003b). "The Dream Lab." *IEEE Potentials*: 8–12.

Jordan, G. B., et al. (2005). "Investigating Differences Among Research Projects and Implications for Managers." *R&D Management* **35** (5): 501–11.

Judge, W. Q., Fryxell, G. E., and Dooley, R. (1997). "The New Task of R&D Management: Creating Goal-Directed Communities for Innovation." *California Management Review* **39** (3): 4–85.

Judson, H. (1979). *The Eighth Day of Creation: The Makers of the Revolution in Biology*. New York: Simon & Schuster.

Kanter, R. M. (1988). "When a Thousand Flowers Bloom: Structural, Collective and Social Conditions for Innovation in Organizations." *Research in Organizational Behavior* 10 (1): 169.

Kao, J. (2007). *Innovation Nation: How America Is Losing Its Innovative Edge, Why It Matters and What We Can Do About It.* New York: Free Press.

Kidder, T. (1981). *The Soul of the New Machine.* New York: Atlantic-Little Brown.

Kim, J., and Wilemon, D. (2007). "The Learning Organization as Facilitator of Complex NPD Projects." *Creativity & Innovation Management* 16 (2): 176–91.

Kintisch, E. (2009). "Projections of Climate Change Go from Bad to Worse." *Science* 323: 1546–47.

Klein, J. (2008). "Evaluation of Interdisciplinary and Transdisciplinary Research: A Literature Review." *Journal of Preventive Medicine* 35 (2S): S116–25.

Klepper, S., and Simons, K. (1997). "Technological Extinctions of Industrial Firms: An Inquiry into Their Nature and Causes." *Industrial and Corporate Change* 6 (2): 379–460.

Kline, S. J., and Rosenberg, N. (1986). "An Overview of Innovation." In R. Landau and N. Rosenberg, eds., *The Positive Sum Strategy: Harnessing Technology for Economic Growth*, 275–306. Washington, DC: National Academy Press.

Kogut, B., Shan, W., and Walker, G. (1993). "Knowledge in the Network and the Network as Knowledge: The Structuring of New Industries." In G. Grabher, ed., *The Embedded Firm: On the Socioeconomics of Industrial Networks*, 67–94. London: Routledge.

Kohler, R. (1994). *Lords of the Fly: Drosophila Genetics and the Experimental Life.* Chicago: University of Chicago Press.

Lafley, A. G., and Charan, R. (2008). *The Game-Changer: How You Can Drive Revenue and Profit Growth with Innovation.* New York: Crown Business.

Lagrange, E. (1954). *Monsieur Roux.* Brussels: Goemaere.

Lawrence, P., and Lorsch, J. (1967). *Organizations and Environment.* Boston: Harvard Business School.

Lazerson, M. (1993). "Factory or Putting Out? Knitting Networks in Modena." In G. Grabher, ed., *The Embedded Firm: On the Socioeconomics of Industrial Networks*, 203–26. New York: Routledge.

Legout, S. (1999). *La Famille Pasteurienne: Le personnel scientifique permanent de l'Institut Pasteur de Paris entre 1889 et 1914.* Paris: School for Advanced Studies in the Social Sciences.

Léonard, J. (1981). *La médecine entre les pouvoirs et les savoirs: Histoire intellectuelle et politique de la médecine française au XIX siècle.* Paris: Aubier Montaigne.

Lépine, P. (1966). *Metchnikoff: Présentation, choix de Textes, bibliographie, portraits, facsimile.* Paris: Seghers.

Levi, M. (1997). "A Model, a Method and a Map: Rational Choice in Comparative and Historical Analysis. In M. Lichbach, and L. Zuckerman, eds., *Comparative Politics: Rationality, Culture, and Structure.* Cambridge: Cambridge University Press.

Levin, D. Z., and Cross, R. (2004). "The Strength of Weak Ties You Can Trust: The Mediating Role of Trust in Effective Knowledge Transfer." *Management Science* **50** (11): 1477–90.

Liebenau, J., and Robson, M. (1991). "L'Institut Pasteur et l'industrie pharmaceutique." In M. Morange, *Centenaire de l'Institut Pasteur*, 52–61. Paris: La Découverte.

Lu, P. H., Edlard, S., et al. (2009). "Donepezil Delays Progression to AD in MCI Subjects with Depressive Symptoms." *Neurology* **72** (24): 2115–21.

Lundvall, B. (1992). *National Systems of Innovation: Towards a Theory of Innovation and Interactive Learning.* London: Pinter.

Malerba, F. (2005). "Sector Systems: How and Why Innovation Differs Across Sectors." In J. Faberberg, D. Mowery, and R. Nelson, eds., *The Oxford Handbook of Innovation*, 380–406. Oxford: Oxford University Press.

Malerba, F., and Orsenigo, L. (1993). "Technological Regimes and Firm Behavior." *Industrial and Corporate Change* **2** (1): 45–72.

Manton, K., et al. (2009). "NIH Funding Trajectories and Their Correlations with U.S. Health Dynamics from 1950 to 2004." *Proceedings of the National Academy Sciences* **106** (27): 10981–86.

Matthews, J. A. (1997). "A Silicon Valley of the East: Creating Taiwan's Semiconductor Industry." *California Management Review* **39** (4): 26–55.

Mendelsohn, A. (1996). *Cultures of Bacteriology: Formation and Transformation of a Science in France and Germany, 1870–1914.* Princeton, NJ: Princeton University Press.

Mery, J. (1987). *Histoire des legs à l'Institut Pasteur: Histoire et Histories.* Paris: Pasteur Institute Archives.

Mohrman, S. A., Galbraith, J. R., and Monge, P. (2004). "Network Attributes Impacting the Generation and Flow of Knowledge Within and from the Basic Science Community." Technical report submitted to the Department of Energy.

Mohrman, S. A., Galbraith, J. R., and Monge, P. R. (2006). "Network Attributes Impacting the Generation and Flow of Knowledge Within and from the Basic Science Community." In J. Hage and M. Meeus, eds., *Innovation, Science, and Institutional Change: A Research Handbook*, 196–216. London: Oxford University Press.

Molas-Gallart, J., and Davies, A. (2006). "Toward Theory-Led Evaluation: The Experience of European Science, Technology, and Innovation Policies." *American Journal of Evaluation* **27** (1): 64–82.

Morton, D., et al. (2006). "Sentinel-Node Biopsy or Nodal Observation in Melanoma." *New England Journal of Medicine* **345** (13): 1307–17.

Mote, J. E., Jordan, G. B., and Hage, J. (2007). "Measuring Radical Innovation in Real Time." *International Journal of Technology, Policy and Management* **7** (4): 355–77.

Moulin, A.-M. (1994). "Bacteriological Research and Medical Practice in and out of the Pastorian School." In A. La Berge and M. Feingold, eds., *French Medical Culture in the Nineteenth Century*, 327–49. Amsterdam: Rodolpi.

Moulin, A.-M., and Guenel, A. (1993). "L'Institut Pasteur et la naissance de l'industrie de la santé." In J.-C. Beaune, ed., *La philosophie du remède*, 91–109. Seyssel, France: Editions Champ Vallon.

National Science Foundation. (2008). *Science and Engineering Indicators*. Washington, DC: National Science Foundation.

National Science Foundation. (2010). *Science and Engineering Indicators*. Washington, DC: National Science Foundation.

Nelson, R. R., ed. (1993). *National Innovation Systems: A Comparative Analysis*. New York: Oxford University Press.

Nieminen, M., and Kaukonen, E. (2001). *Universities and R&D Networking in a Knowledge-Based Economy: A Glance at Finnish Developments*. Sitra Report Series 11. Helsinki: Sitra.

Nonaka, I., and Peltokorpi, V. (2006). "Knowledge-Based View of Radical Innovation: Toyota Prius Case." In J. Hage and M.T.H. Meeus, eds., *Innovation, Science, and Institutional Change: A Research Handbook*, 88–104. Oxford: Oxford University Press.

Nonaka, I., and Takeuchi, H. (1995). *The Knowledge-Creating Company: How Japanese Companies Create the Dynamics of Innovation*. Oxford: Oxford University Press.

Nooteboom, B. (1999). *Inter-Firm Alliances: Analysis & Design*. London: Routledge.

Nooteboom, B., et al. (2007). "Optimal Cognitive Distance and Absorptive Capacity." *Research Policy* **36** (7): 1016–34.

Normile, D. (2009). "Persevering Researchers Make a Splash with Farm-Bred Tuna." *Science* **324**: 1260–61.

O'Doherty, D. (1995). *Globalisation, Networking, and Small Firm Innovation*. London: Graham and Trotman.

Ohlrogge, J., et al. (2009). "Driving on Business." *Science* **324**: 1019–20.

Pasteur Institute. (1911). *Rapport des chiefs de services de l'Institut Pasteur*. Paris: Cour d'Appel.

Pavitt, K. (1984). "Sectoral Patterns of Technical Change: Towards a Taxonomy and a Theory." *Research Policy* **13** (6): 343–73.

Pelis, K. (1995). *Pasteur's Imperial Missionary: Charles Nicolle, 1866–1936, and the Pasteur Institute of Tunis*. Baltimore: Johns Hopkins University Press.

Pelz, D., and Andrews, F. M. (1976). *Scientists in Organizations: Productive Climates Research and Development*. Ann Arbor: University of Michigan Press.

Penn, M., and Zalesne, E. (2007). *Microtrends: The Small Forces Behind Tomorrow's Big Changes*. New York: Twelve Hachette Group.

Pernish, E. (2009). "Darwin Applies to Medical School." *Science* **323**: 1004–5.

Pfeffer, J. (1981). *Power in Organizations*. Boston: Pitman.

Pierson, P. (2000). "Increasing Return, Path Dependence, and the Study of Politics." *American Political Science Review* **94**: 251–67.

Pine, B. J. (1993). *Mass Customization: The New Frontier in Business Competition*. Cambridge: Harvard Business Press.

Piore, M. J., and Sabel, C. F. (1984). *The Second Industrial Divide: Possibilities for Prosperity*. New York: Basic Books.

Polanyi, M. (1966). *The Tacit Dimension*. London: Routledge.

Powell, W. W. (1998). "Learning from Collaboration: Knowledge and Networks in the Biotechnology and Pharmaceutical Industries." *California Management Review* **40** (3): 228–40.

Powell, W., and Brantley, P. (1992). "Competitive Cooperation in Biotechnology: Learning Through Networks?" In N. Nohria, and R. Eccles, eds., *Networks and Organizations: Structure, Form, and Action*, 365–94. Boston: Harvard Business School Press.

Powell, W. W., Koput, K. W., and Smith-Doerr, L. (1996). "Interorganizational Collaboration and the Locus of Innovation: Networks of Learning in Biotechnology." *Administrative Science Quarterly* **41** (1): 116–45.

Prais, S., Jarvin, S., and Wagner, K. (1989). "Productivity and Vocational Skills in Britain and Germany: Hotels." *National Institute Economic Review* (November).

Prais, S., and Steedman, H. (1986). "Vocational Training in France and Britain in the Building Trades." *National Institute Economic Review* **116**: 45–56.

Putnam, R., et al. (1993). *Making Democracy Work: Civic Traditions in Modern Italy.* Princeton, NJ: Princeton University Press.

Rafols, I., and Meyer, M. (2010). "Diversity and Network Coherence as Indicators of Interdisciplinarity: Case Studies in Bionanoscience." *Scientometrics* **82**: 263–86.

Rassmussen, W. (1989). *Taking the University to the People: Seventy-Five Years of Cooperative Extension.* Ames: University of Iowa Press.

Robinson, M. (1906). *La vie d'Emile Duclaux.* Laval, FR: L. Barneoud et Cie.

Romer, P. (1994). "The Origins of Endogenous Growth." *Journal of Economic Perspectives* **8** (1): 3–22.

Sánchez, A., and Pérez, N. (2003). "Supply Chain Flexibility and Firm Performance: A Conceptual Model and Empirical Study in the Automobile Industry." *International Journal of Operations and Production Management* **25** (7): 681–700.

Sawyer, K. (2007). *Group Genius: The Creative Power of Collaboration.* New York: Basic Books.

Seidell, A. (1926). "Chemical Research at the Pasteur Institute." *Journal of Chemical Education* **3**: 1217–39.

Seria, R. (2009). "Hydrogen Cars: Fads or Future?" *Science* **324**: 1257–59.

Shaiken, H. (1984). *Work Transformed.* New York: Holt, Rinehart and Winston.

Shenhar, A. (1999). "Systems Engineering Management: The Multidisciplinary Discipline." In A. Sage, ed., *Handbook of Systems Engineering*, 394–414. New York: Wiley.

Shenhar, A. J. (2001). "One Size Does Not Fit All Projects: Exploring Classical Contingency Domains." *Management Science* **47** (3): 394–414.

Sicotte, H., and Langley, A. (2000). "Integration Mechanisms and R&D Project Performance." *Journal of Engineering & Technology Management* **17** (1): 1–37.

Steedman, H., and Wagner, K. (1987). "A Second Look at Productivity: Machinery and Skills in Britain and Germany. *National Institute Economic Review* **122**: 84–95.

Steedman, H., and Wagner, K. (1989). "Productivity, Machinery and Skills: Clothing Manufacture in Britain and Germany." *National Institute Economic Review* **128**: 40–57.

Stokes, D. E. (1997). *Pasteur's Quadrant.* Washington, DC: Brookings Institution Press.

Stokols, D., et al. (2008). "The Science of Team Science: Overview of the Field and Introduction to the Supplement." *American Journal of Preventive Medicine* **35** (2S): S77–89.

Streeck, W., et al. (1987). *Steuerang und Regulierung der Beruflichen Bildung: die Rolle der Sozialpartner in der Ausbildung und Heruflichen Weiterbildung in der Br Deutschland.* Berlin: Sigma.

Tassey, G. (2009). Rationales and Mechanisms for Revitalizing U.S. Manufacturing R&D Strategies. Paper. National Institute of Standards and Technology. Gaithersburg, MD.

Utterback, J. (1994). *Mastering the Dynamics of Innovation: How Companies Can Seize Opportunities in the Face of Technological Change.* Boston: Harvard Business School.

van Waarden, B. F., and Oosterwijk, H. G. M. (2006). "Turning Tracks? Path Dependency, Technological Paradigm Shifts, and Organizational and Institutional Change." In J. Hage and M. T. H. Meeus, eds., *Innovation, Science, and Institutional Change: A Research Handbook,* 443–64. London: Oxford University Press.

van Waarden, B. F., Unger, B., et al. (2003). National Systems of Innovation and Networks in the Idea-Innovation Chain in Science-Based Industries. Report. European Commission.

Walton, R. (1987). *Innovating to Compete: Lessons for Diffusing and Managing Change.* San Francisco: Jossey-Bass.

Weindling, P. (1991). "Emile Roux et la dipthérie." In M. Morange, ed., *L'Institut Pasteur: Contribution à son histoire,* 137–45. Paris: La Découverte.

Weisz, G. (1995). *The Medical Mandarins: The French Academy of Medicine in the Nineteenth and Early Twentieth Centuries.* Oxford: Oxford University Press.

Whitley, R. (1992). *Business Systems in East Asia: Firms and Markets in Their National Context.* London: Sage.

Whitley, R., ed. (1994). *European Business Systems: Firms and Markets in Their National Contexts.* London: Sage.

Womack, J. P., Jones, D. T., and Roos, D. (1990). *The Machine That Changed the World.* New York: Rawson.

Woodman, R., Sawyer, J., and Griffin, R. (1993). "Toward a Theory of Organizational Creativity." *Academy of Management Review* **18** (2) (April): 293–321.

Wuchty, S., Jones, B. F., and Uzzi, B. (2007). "The Increasing Dominance of Teams in the Production of Knowledge." *Science* **316**: 1036–39.

Wuyts, S., Colombo, M. G., et al. (2005). "Empirical Tests of Optimal Cognitive Distance." *Journal of Economic Behavior & Organization* **58** (2): 277–302.

Zammuto, R. F., and O'Connor, E. J. (1992). "Gaining Advanced Manufacturing Technologies' Benefits: The Roles of Organization Design and Culture." *Academy of Management Review* **17** (4): 701–28.

INDEX

ABOUT THE AUTHOR

Jerald Hage has been Director, Center for Innovation, University of Maryland in College Park, since 1982. He started studying organizational innovation in the 1960s and since then has also worked on institutional analysis in health, education, and welfare, primarily with comparative studies of Europe. For this reason, he was elected president of the Society for the Advancement of Socio-Economics in 1998. Another major interest has been in theory construction and, in particular, the development of criteria for deciding when a concept, model, or theory is useful. Besides this book, he has authored or co-authored sixteen books in these areas as well as written over one hundred papers. In the last decade and a half, he has been concentrating his attention on science and technology, especially the development of the science of science and innovation theory, to which this book contributes. Presently he is directing two major research projects in these areas that are funded by the National Science Foundation as well as a project for the Center for Satellite Applications and Research (STAR) of the National Oceanographic and Atmospheric Administration.